Pharmaceutical Engineering

Practical Manual

(Unit Operations)

Second Edition

Pharmaceutical Engineering
Practical Manual
(Unit Operations)
Second Edition

Sudhakara Reddy Pondugula
M.pharm
Alphamed Formulations Pvt. Ltd.,
Survey No. 225, Sampanbole Village,
Shamirpet, R.R.Dist Hyderabad.

G.V. Chandrasekhar
M.pharm
Research Scientist
APLRC, Unit-IV a division of
Aurobindo Pharma Ltd.,
Miyapur, Hyderabad.

M. Sharath Chandra
M.pharm
Asst. Professor, PNR College of Pharmacy &
Managing Director, Elixer Academy,
Hyderabad.

Pharma Med PRESS
An Imprint of
BSP Books Pvt. Ltd.

PharmaMed Press
An imprint of BSP Books Pvt. Ltd.
4-4-309/316, Giriraj Lane,
Sultan Bazar, Hyderabad - 500 095.

First Edition, 2007
Second Edition, 2009

Eighth Reprint 2022

Published by

Pharma Med PRESS
An Imprint of
BSP Books Pvt. Ltd.

PharmaMed Press
An imprint of BSP Books Pvt. Ltd.
4-4-309/316, Giriraj Lane, Sultan Bazar, Hyderabad - 500 095.
Phone: 040-23445688; Fax: 91+40-23445611
E-mail: info@pharmamedpress.com
www.pharmamedpress.com/pharmamedpress.net

Printed at

Repro India Limited.

Price: Rs. 225.00

ISBN: 978-81-88449-87-3 (Paperback)

Foreword

Prof. M. Sadanandam
M.S., Ph.D
Principal, Sri Venkateswara College of Pharmacy
Madhapur, Hyderabad.

I am very much happy to express myself about this book and authors. This book consists of 31 experiments and each experiment consists of theory part as well as experimental procedures. All experiments were explained diagrammatically. This book is simple and easy to understand for the staff as well as students. This book also contains DEFINITIONS and VIVA VOCE questions for each topic which are incorporated.

Regarding authors, two of these were graduated from our college only and completed their post graduation from Sri Rama Krishna College of Pharmaceutical Sciences, Coimbatore. It is one of the top colleges in Tamil Nadu. They are young talented, their amiable group work is appreciable.

Dr. M. SADANANDAM

Preface to Second Edition

The need has arisen to revise the first edition of the book with the inclusion of the following four experiments:

1. Organised Substances by Percolation Method
2. Continuous Hot Extraction Method
3. Factors Affecting the Rate of Evaporation
4. Effect of Particle Size on Sedimentation Rate

As usual, we maintained lucid language and self-explanatory form. We are sure, with these additions, the students now can cover the subject much more.

The authors are thankful to Mr.Anil Shah, PharmaMed Press for bringing the new edition in record time.

Constructive suggestions are welcome, which will help us for increasing the experiments.

-Authors

Preface to First Edition

It gives us immense pleasure in bringing out Pharmaceutical Engineering Practical Manual.

The need for the manual was felt necessary, as it is difficult to conduct experiments without specific laboratory manual. We approached some of our friends and well-wishers regarding this project. They gave suggestions and extended their contribution to complete this manual. We hope, our colleagues, who are handling this subject in their institutions feel the same.

The presentations of fundamental concepts are in lucid and self-explanatory form. The unique feature of this book is that it contains 130 definitions and numbers of viva questions for each experiment based on topic. This manual can be used as observations notebook for which space is provided at the end of each experiment. The students can acquire more knowledge with the observations.

We welcome suggestions and criticism to improve the manual by lecturers and students.

Pharmaceutical Engineering is called Unit operations in other Universities.

-Authors

Acknowledgment

We extend our sincere thanks to the following Principals for their encouragement to complete this project successfully.

1. Prof. Dr. K.V. Ramana Murthy, Dept. of Pharmaceutics, Andhra University, Vishakapatnam.
2. Mr. Rambabu, Manager, Alphamed Formulations Pvt Ltd. Hyderabad.
3. Prof. M. Sadanandam, Principal, Sri Venkateswara College of Pharmaceutical Sciences, Hyderabad. We are also grateful to him for providing Foreword to this book.
4. Dr.G. Sinivas Reddy; Secretary & Correspondent, Vaageswari Educational Society.
5. Prof. C.V.S. Subrahmanyam, Principal, Gokaraju Rangaraju College of Pharmaceutical Sciences, Hyderabad.
6. Dr. Dayakar, Principal, Blue Bird College of Pharmacy.
7. Dr. Gampa Vijay Kumar, Principal, Vikas Institute of Pharmaceutical Sciences.
8. Dr. Babu Rao, Principal, Hindu College of Pharmacy
9. Dr. Ravi Kumar, Principal, Priyadarshini College of Pharmacy

Our thanks to all our colleagues for their interaction during the preparation of the manuscript, Mr.Dinesh S. Gujarathi, Associate Professor, Mr. Digpathi Roy, Professor, Mr. Satyanarayana, Research Scholar (AU), Mr. Ramalingam, Associate Professor, Mr. P.Praveen Kumar, Asst.Professor, MR.Rayees Ahmed, Asst.Professor, Mr. Nyamthulla, Research Scholar (AU), Mr. Ravi Kumar, Research Scholar (AU), Mr. S.Suresh, Asst.Professor, Mr. Mahipal Reddy, Associate Professor, Mr. Venkata Ramana, Asst.Professor.

We specially thank Mrs. P. Pallavi, Research Associate, Mrs. G. Samatha, Asst. Professor for their valuable guidance to complete this book.

We also thank various Pharmacy College Principals & Lecturers across the nation for their encouragement towards our manual.

- Authors

Contents

Symbols

ρ	=	Density (lb mass / cu. ft.)
ε	=	Emissivity
μ, η	=	Viscosity (lb mass / ft. sec)
π	=	Constant (22/7 = 3.14)
β	=	Coefficient of thermal expansion; angle
θ	=	Time (sec)
Δ	=	Finite difference (Difference in)
λ	=	Latent heat of Vaporization
g	=	Acceleration of gravity (ft/sec^2)
r	=	Radius
d	=	Diameter
u	=	Average velocity
R_e	=	Reynolds Number
f	=	Friction factor
A	=	Area (sq. ft.)

Stefan Boltzman Equation

$$\mathbf{Q_R = H \times t}$$

Where H = Heat transfer coefficient

t = time in sec.

$$\mathbf{H = \varepsilon.\sigma.A\,(T_1^{\,4} - T_2^{\,4})}$$

Radiation Constant (σ) value in	**W/m^2 ^{o}C**	**Btu/hr. Sq. ft. ^{o}F^4**
For Brass -	0.152	0.036
For Iron -	5.09	0.154
For Copper -	0.119	0.028
For Glass -	5.13	0.155

Important Equations Helpful in Practicals

Vernier Calipers :

Least Count (L.C.) : $\frac{S}{N} = \frac{\text{Main scale division}}{\text{No.of divisions on Vernier scale}}$ cm

Radius or length of a cylinder : [MSR + (VCD × L.C.)]

MSR = Main scale reading; VCD = Vernier coincide divisions

$$\text{Density of solid} = \frac{\text{Wt. of solid in air}}{\text{Loss of Wt. in water}}$$

$$\text{Specific gravity of Solid} = \frac{\text{Wt. of certain equal volume of Liquids}}{\text{Wt. of equal volume of Liquid}}$$

Volume of Cylinder = (V) = $\pi r^2 l$ cm^3

Volume of Sphere = (Vs) = 4/3 πr^3 cm^3

Screw gauge :

Least count (L.C.) = $\frac{\text{Pitch of the screw}}{\text{No.of head scale divisions}}$ mm

Diameter of Sphere = [PSR + (HSR × L.C.)]

PSR = Pitch scale reading; HSR = Head scale reading

Circumference of Circle = 2 πr^2 cm^2

Area of Circle = πr^2 cm^2

Volume of Circle = $\pi r^2 h$ cm^3

Area of Cylinder = $2\pi r$ (h + R) cm^2

Surface area of Sphere = $4\pi r^2$ cm^2

Area of curved surface of cone = πrl cm^2

$$^oF = 9/5\ ^oC + 32$$

$$^oR = {}^oF + 459.4$$

Velocity = Volume / (Area × Time)

Experiment No : 01 **Dt :**

Determination of Radiation Constant of Iron Cylinder

Aim : To determine the radiation constant of Iron cylinder.

Requirements : Iron cylinder, Thermometer, Wooden plank, Stop clock, Tongs & heating source.

Theory : Radiation is a term given to the transfer of energy through space by means of electromagnetic waves. All solid bodies with a temperature above absolute zero, radiates energy. If radiation is passing through empty space, it is not transformed to heat or any other form of energy and it is not diverted from its path. If, however, matter appears in its path, the radiation will be transmitted, reflected, or absorbed. It is only the absorbed energy that appears as heat, and this transformation is quantitative. For example, fused quartz transmits practically all the radiation that strikes it; a polished opaque surface or mirror will reflect most of the radiation impinging on it; a black or a matte surface will absorb most of the radiation received by it and will transform such absorbed energy quantitatively into heat.

Heat transfer by radiation is known as *Thermal Radiation.* Heat transfer (thermal energy) is predominant as the temperature of the body increases. The amount and kind of thermal energy radiated increases rapidly with temperature by *Convection* and *Conduction.*

Conduction : "Heat flow through a body, by transference of momentum of individual atoms or molecules without mixing". This mechanism is practically restricted to solids. It is rare in fluids since eddies are formed due to varying local densities.

Principle : Heat is lost from hot cylinder surface to surrounding atmosphere by means of '**Conduction**' and '**Radiation**'. Heat transfer by radiation occurs, energy transfer through space by means of electromagnetic radiation (waves). Thus a body acts as an emitter, than energy being transmitted through the intervening spaces. It is effective even in a perfect vacuum or inter-spacious space. The amount of thermal energy radiated by a surface is increased rapidly with increasing temperature, when heat flows by actual mixing of warmer portions with cooler portions of the same material. This mechanism is known as **"Convection"**.

The rate of heat transfer by a hot body equal to sum of heat transfer through conduction and radiation

$$\boxed{Q = Q_C + Q_R} \qquad .....(1)$$

Where, **Q** = Rate of heat loss/transfer in **Btu/hr**

Q_C = Rate of heat transfer by conduction in **Btu/hr**

Q_R = Rate of heat transfer by radiation in **Btu/hr**

The rate of heat transfer through a heat body to 0 °C per time

$$\boxed{Q = MC_P\, d\theta/dt} \qquad .....(2)$$

Where, **M** = **Mass** of the cylinder in gm.

C_p = **Specific heat** of material (Cylinder) in Cal/gm/°C

dθ/dt = **Slope** of the graph plotted between time on X-axis and temperature on Y- Axis

Q = Rate of heat transfer in **Btu/hr**

The rate of heat transfer through radiation is

$$\boxed{Q_R = \varepsilon A \sigma (T_1^4 - T_2^4)} \qquad .....(3)$$

(***Stefan–Boltzmann's Law*** gives the total amount of radiation emitted by a body)

Where,

- ε = Emissivity of the cylinder (for black body ε = 1)
- A= Area of the cylinder [$2\pi rh + \pi r^2$] in Sq.cm.
- σ = Stephan–Boltzman constant (for black bodies b = 0.174×10^{-8} Btu/hr. Sq. ft. $°F^4$ or 5.67×10^{-8} $W/m^2.K^4$)
- T_1 & T_2 are the absolute temperatures of initial & final in °C respectively.

Heat loss through conducts :

$$\boxed{Q_c = \beta \left[\frac{T_1 - T_2}{D}\right]^{0.25}} \qquad .....(4)$$

β = Convection Constant (Iron = 0.28)

D = Diameter of cylinder

Procedure

1. Select an Iron cylinder whose surface is smooth & radiation constant is determined.
2. Note & record the room temperature, measure the weight, area and radius of the cylinder. Insert the thermometer into the cavity of the cylinder.
3. Place the cylinder on a tripod stand and heat it for about 300 °C
4. Now hold the cylinder with the tongs and place on a non-conducting surface (wooden plank), without touching any surface.
5. Note the temperature reading for every 5 min, by using stop clock.
6. Plot a graph between 'temperature' on Y-axis and 'time' on X-axis.
7. Find out the slopes $d\theta/dt$, at various arbitrary temperatures.
8. Calculate the radiation constant by using equation.

Precautions

1. The cylinder must be transferred as fast as possible.
2. Avoid parallax error while recording the temperature.

Observation

S. No.	Time (Min)	Temp. (°C)
01	0	Above 300°C
02	5	
03	10	
04	15	
05	20	
06	25	
07	30	
08	35	
09	40	
10	45	
11	50	
12	55	
13	60	
14	65	
15	70	

(m) Wt. of cylinder = gm.
(H) Height of cylinder = cm.
(D) Diameter = cm.

$$\text{Area (A)} = 2\pi rh + \pi r^2 \text{ in Sq.cm}$$

$$^{o}F = 9/5\,^{o}C + 32 \text{ [For } T_1 \text{ \& } T_2]$$

$$^{o}R = {}^{o}F + 459.4$$

$$Q = M.\ C_p.\ d\theta/dt \quad [d\theta/dt = (Y_2 - Y_1)/(X_2 - X1)]$$

C_p = Specific heat of Iron = 0.44 J/g. K.

$$Q_R = Q - Q_C \text{ (in Btu/hr)}$$

$$Q_c = \beta \left(\frac{T_1 - T_2}{D} \right)^{0.25}$$

$\sigma = Q_R / A.\varepsilon\,(T_1^4 - T_2^4)$ (in Btu/hr. Sq.cm. ^{o}F) for iron – $\varepsilon = 0.65$

Report :

Graphs : Graphical representation

(a) Determination of slope

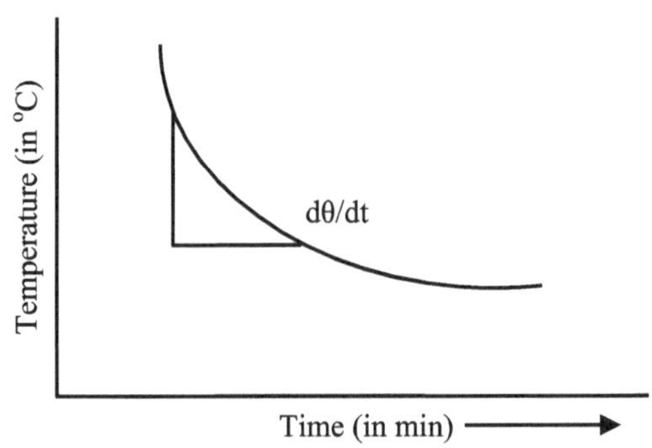

(b) Influence of temperature on radiation constant

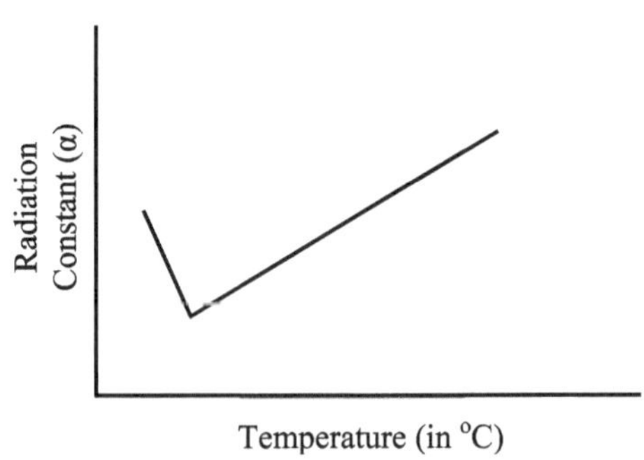

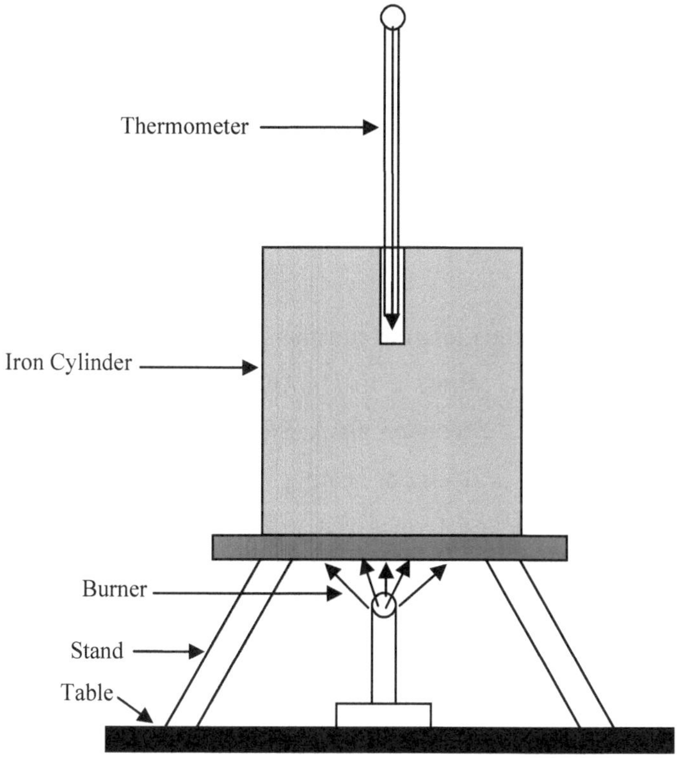

Fig. 1.1 Experimental setup for Determination of Radiation constant of Iron cylinder.

Bibliography

1. Introduction to chemical engineering, 7th Edition; 2002, by **Walter L. Badger and Juliust T. Banchero** page No. 117 & 152 – 155.
2. Pharmaceutical Engineering principles and practices, 6th Edition, by **C.V.S. Subrahmanyam**., page No. 105 – 130.
3. Pharmaceutical Engineering, 2003 1st Edition, by **K. Sambamurthy,** page No. 75 – 77.
4. Cooper & Gun's, Tutorial Pharmacy, 6th Edition, Edited by **S.J. Carter**., Page No. 149 – 150.
5. Unit operations of chemical engineering, 6th Edition, by **M.C. Cabe, J.C.Smith, harriott,** page no. 425 – 426.

Experiment No : 02 **Dt :**

Determination of Radiation Constant of Brass Cylinder

Aim : To determine the radiation constant of Brass cylinder.

Requirements : Brass cylinder, Thermometer,
Wooden plank, Stop clock,
Tongs & heating source

Theory : Radiation is a term given to the transfer of energy through space by means of electromagnetic waves. All solid bodies with a temperature above absolute zero, radiates energy. If radiation is passing through empty space, it is not transformed to heat or any other form of energy and it is not diverted from its path. If, however, matter appears in its path, the radiation will be transmitted, reflected, or absorbed. It is only the absorbed energy that appears as heat, and this transformation is quantitative. For example, fused quartz transmits practically all the radiation that strikes it; a polished opaque surface or mirror will reflect most of the radiation impinging on it; a black or a matte surface will absorb most of the radiation received by it and will transform such absorbed energy quantitatively into heat.

Heat transfer by radiation is known as *Thermal Radiation.* Heat transfer (thermal energy) is predominant as the temperature of the body increases. The amount and kind of thermal energy radiated increases rapidly with temperature by *Convection* and *Conduction.*

Conduction : "Heat flow through a body by transference of momentum of individual atoms or molecules without mixing". This mechanism is practically restricted to solids. It is rare in fluids since eddies are formed due to varying local densities.

Principle : Heat is lost from hot cylinder surface to surrounding atmosphere by means of '**Conduction**' and '**Radiation**'. Heat transfer by radiation occurs, energy transfer through space by means of electromagnetic radiation (waves). Thus a body acts as an emitter, than energy being transmitted through the intervening spaces. It is effective even in a perfect vacuum or inter-spacious space. The amount of thermal energy radiated by a surface is increased rapidly with increasing temperature, when heat flows by actual mixing of warmer portions with cooler portions of the same material. This mechanism is known as "**Convection**".

The rate of heat transfer by a hot body equal to sum of heat transfer through conduction and radiation

$$\mathbf{Q = Q_C + Q_R} \qquad .....(1)$$

Where, **Q** = Rate of heat loss/transfer in **Btu/hr**

$\mathbf{Q_C}$ = Rate of heat transfer by conduction in **Btu/hr**

$\mathbf{Q_R}$ = Rate of heat transfer by radiation in **Btu/hr**

The rate of heat transfer through a heat body to 0 °C per time

$$\mathbf{Q = MC_P\ d\theta/dt} \qquad .....(2)$$

Where, **M** = **Mass** of the cylinder in gm.

$\mathbf{C_p}$ = **Specific heat** of material (Cylinder) in Cal/gm/°C

dθ/dt = **Slope** of the graph plotted between time on X-axis and temperature on Y- Axis

Q = Rate of heat transfer in **Btu/hr**

The rate of heat transfer through radiation is

$$\mathbf{Q_R = \varepsilon A\sigma\ (T_1^4 - T_2^4)} \qquad .....(3)$$

(***Stefan–Boltzmann's Law*** gives the total amount of radiation emitted by a body)

Where,

- ε = Emissivity of the cylinder (for black body ε = 1)
- A = Area of the cylinder $[2\pi rh + \pi r^2]$ in Sq.cm.
- σ = Stephan–Boltzman constant (for black bodies b = 0.174 × 10^{-8} Btu/hr. Sq. ft. $°F^4$)
- T_1 & T_2 are the absolute temperatures of initial & final in °C respectively.

Heat loss through conducts:

$$\mathbf{Q_c = \beta \left[\frac{T_1 - T_2}{D}\right]^{0.25}} \qquad .....(4)$$

β = Convection Constant (Brass = 0.28)

D = Diameter of cylinder

Procedure :

1. Select a Brass cylinder whose surface is smooth & radiation constant is determined.
2. Note & record the room temperature, measure the weight, area and radius of the cylinder. Insert the thermometer into the cavity of the cylinder.
3. Place the cylinder on a tripod stand and heat it for about 300°C
4. Now hold the cylinder with the tongs and place on a non-conducting surface (wooden plank), without touching any surface.
5. Note the temperature reading for every 5 min, by using stop clock.
6. Plot a graph between 'temperature' on Y-axis and 'time' on X-axis.
7. Find out the slopes $d\theta/dt$, at various arbitrary temperatures.
8. Calculate the radiation constant by using equation.

Precautions :

1. The cylinder must be transferred as fast as possible.
2. Avoid parallax error while recording the temperature.

Observation :

S. No.	Time (Min)	Temp. (°C)
01	0	Above 300°C
02	5	
03	10	
04	15	
05	20	
06	25	
07	30	
08	35	
09	40	
10	45	
11	50	
12	55	
13	60	
14	65	
15	70	

(m) Wt. of cylinder = **gm.**

(H) Height of cylinder = **cm.**

(D) Diameter = **cm.**

Area (A) = $2\pi rh + \pi r^2$ in Sq.cm.

$^oF = 9/5\,^oC + 32$ [For T_1 & T_2]

$^oR = {^oF} + 459.4$

$Q = M.\ C_p.\ d\theta/dt$ **[$d\theta/dt = (Y_2 - Y_1)/(X_2 - X1)$]**

C_p = Specific heat of Brass = 0.092 Btu/lb. F

$Q_R = Q - Q_C$ (in Btu/hr)

$$Q_c = \beta \left(\frac{T_1 - T_2}{D}\right)^{0.25}$$

$\sigma = Q_R / A.\varepsilon\,(T_1^4 - T_2^4)$ (in **Btu/hr. Sq.cm. $^oF^4$**) for Brass – $\varepsilon = 0.22$

Report :

Graphs : Graphical representation

(a) Determination of slope

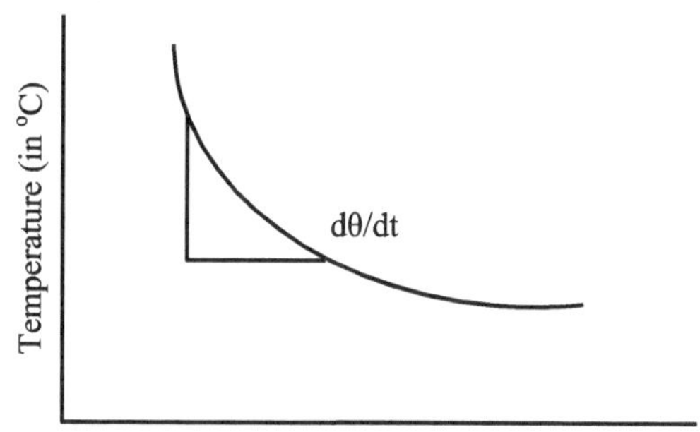

Time (in min) ⟶

(b) Influence of temperature on radiation constant

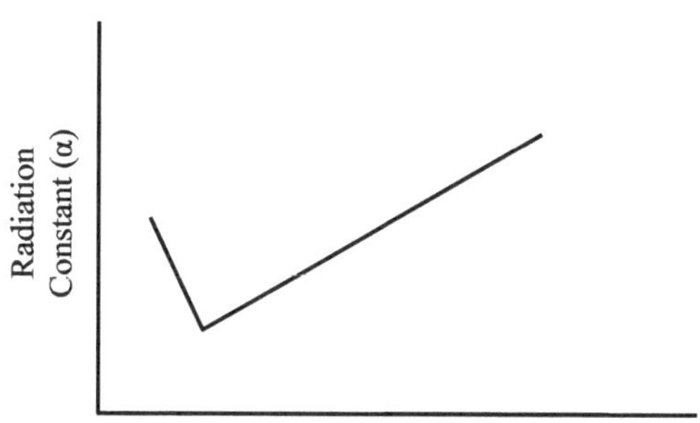

Temperature (in °C)

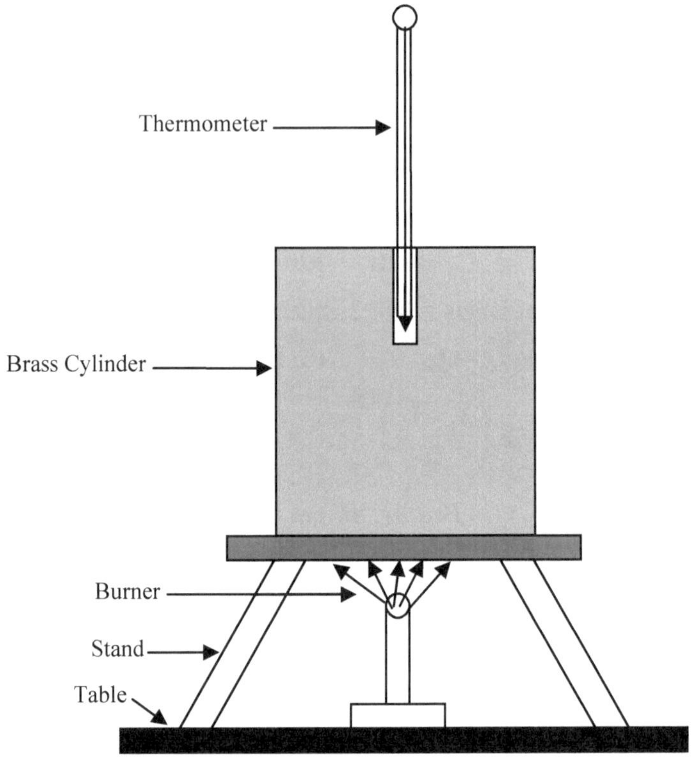

Fig. 2.1 Experimental setup for Determination of Radiation constant of Brass cylinder.

Bibliography :

1. Introduction to chemical engineering, 6th Edition, by **Walter L. Badger and Juliust T. Banchero** page No. 117 & 152 – 155.
2. Pharmaceutical Engineering principles and practices, 6th Edition, by **C.V.S. Subrahmanyam**., page No. 105 – 130.
3. Pharmaceutical Engineering, 2003 1st Edition, by **K. Sambamurthy,** page No. 75 – 77.
4. Cooper & Gun's, Tutorial Pharmacy, 6th Edition, Edited by **S.J. Carter**., Page No. 149 – 150.
5. Unit operations of chemical engineering, 6th Edition, by **M.C. Cabe, J.C.Smith, harriott,** page no. 425 – 426.

Experiment No : 03 **Dt :**

Determination of Radiation Constant of Copper Cylinder

Aim : To determine the radiation constant of Copper cylinder

Requirements : Copper cylinder, Thermometer,
Wooden plank, Stop clock,
Tongs & heating source

Theory : Radiation is a term given to the transfer of energy through space by means of electromagnetic waves. All solid bodies with a temperature above absolute zero, radiates energy. If radiation is passing through empty space, it is not transformed to heat or any other form of energy and it is not diverted from its path. If, however, matter appears in its path, the radiation will be transmitted, reflected, or absorbed. It is only the absorbed energy that appears as heat, and this transformation is quantitative. For example, fused quartz transmits practically all the radiation that strikes it; a polished opaque surface or mirror will reflect most of the radiation impinging on it; a black or a matte surface will absorb most of the radiation received by it and will transform such absorbed energy quantitatively into heat.

Heat transfer by radiation is known as *Thermal Radiation.* Heat transfer (thermal energy) is predominant as the temperature of the body increases. The amount and kind of thermal energy radiated increases rapidly with temperature by *Convection* and *Conduction.*

Conduction : "Heat flow through a body by transference of momentum of individual atoms or molecules without mixing". This mechanism is practically restricted to solids. It is rare in fluids since eddies are formed due to varying local densities.

Principle : Heat is lost from hot cylinder surface to surrounding atmosphere by means of '**Conduction**' and '**Radiation**'. Heat transfer by radiation occurs, energy transfer through space by means of electromagnetic radiation (waves). Thus a body acts as an emitter, than energy being transmitted through the intervening spaces. It is effective even in a perfect vacuum or inter-spacious space. The amount of thermal energy radiated by a surface is increased rapidly with increasing temperature, when heat flows by actual mixing of warmer portions with cooler portions of the same material. This mechanism is known as "**Convection**".

The rate of heat transfer by a hot body equal to sum of heat transfer through conduction and radiation

$$Q = Q_C + Q_R \qquad .....(1)$$

Where, **Q** = Rate of heat loss/transfer in **Btu/hr**

Q_C= Rate of heat transfer by conduction in **Btu/hr**

Q_R = Rate of heat transfer by radiation in **Btu/hr**

The rate of heat transfer through a heat body to 0 °C per time

$$Q = MC_P\, d\theta/dt \qquad .....(2)$$

Where, **M** = **Mass** of the cylinder in gm.

C_p = **Specific heat** of material (Cylinder) in Cal/gm/°C

dθ/dt = **Slope** of the graph plotted between time on X-axis and temperature on Y-Axis

Q = Rate of heat transfer in **Btu/hr**

The rate of heat transfer through radiation is

$$Q_R = \varepsilon A \sigma\,(T_1^4 - T_2^4) \qquad .....(3)$$

(***Stefan–Boltzmann's Law*** gives the total amount of radiation emitted by a body)

Where,

- ε = Emissivity of the cylinder (for black body ε = 1)
- A = Area of the cylinder [$2\pi rh + \pi r^2$] in Sq.cm.
- σ- Stephan–Boltzman constant (for black bodies b = 0.174×10^{-8} Btu/hr. Sq. ft. °F^4)
- T_1 & T_2 = are the absolute temperatures of initial & final in °C respectively.

Heat loss through conducts :

$$Q_c = \beta \left[\frac{T_1 - T_2}{D}\right]^{0.25} \qquad .....(4)$$

Convection Constant (Copper = 0.28)

D = Diameter of cylinder

Procedure :

1. Select a Copper cylinder whose surface is smooth & radiation constant is determined.
2. Note & record the room temperature, measure the weight, area and radius of the cylinder. Insert the thermometer into the cavity of the cylinder.
3. Place the cylinder on a tripod stand and heat it for about 300 °C
4. Now hold the cylinder with the tongs and place on a non-conducting surface (wooden plank), without touching any surface.
5. Note the temperature reading for every 5 min, by using stop clock.
6. Plot a graph between 'temperature' on Y-axis and 'time' on X-axis.
7. Find out the slopes $d\theta/dt$, at various arbitrary temperatures.
8. Calculate the radiation constant by using equation.

Precautions :

1. The cylinder must be transferred as fast as possible.
2. Avoid parallax error while recording the temperature.

Observation :

S. No.	Time (Min)	Temp. (°C)
01	0	Above 300°C
02	5	
03	10	
04	15	
05	20	
06	25	
07	30	
08	35	
09	40	
10	45	
11	50	
12	55	
13	60	
14	65	
15	70	

(m) Wt. of cylinder = gm.
(H) Height of cylinder = cm.
(D) Diameter = cm.

Area (A) = $2\pi rh + \pi r^2$ in Sq.cm.

$^oF = 9/5\,^oC + 32$ [For T_1 & T_2]

$^oR = ^oF + 459.4$

$Q = M.\ C_p.\ d\theta/dt$ **[$d\theta/dt$** = $(Y_2 - Y_1)/(X_2 - X1)$**]**

C_p = Specific heat of Copper = 0.0923 Btu/lb. F

$Q_R = Q - Q_C$ (in Btu/hr)

$$Q_c = \beta \left[\frac{T_1 - T_2}{D}\right]^{0.25}$$

$Q_R / A.\varepsilon\ (T_1^4 - T_2^4)$

(in **Btu/hr. Sq. cm. $^oF^4$**) for Copper – $\varepsilon = 0.78$

Report :

Graphs : Graphical representation

(a) Determination of slope

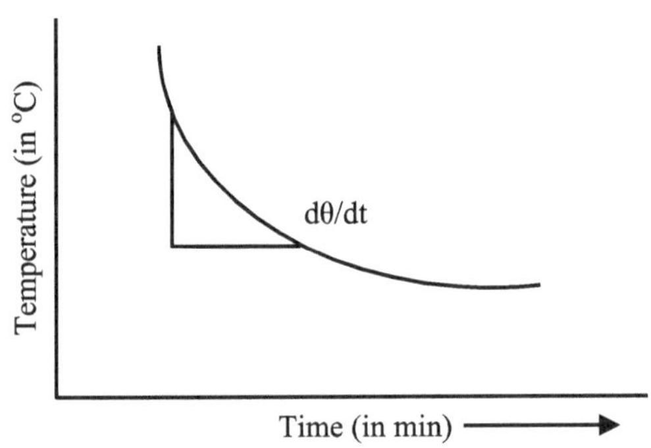

(b) Influence of temperature on radiation constant

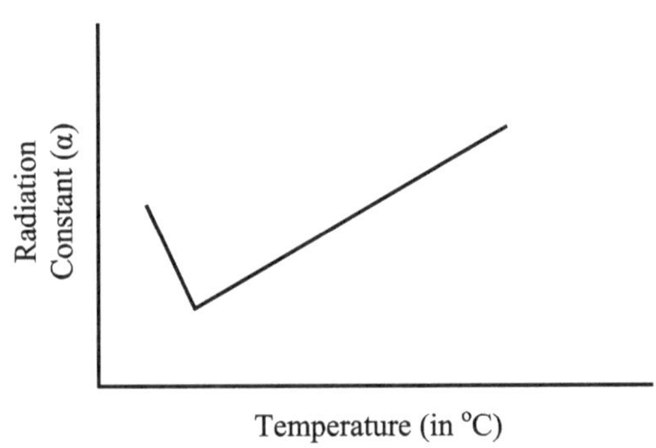

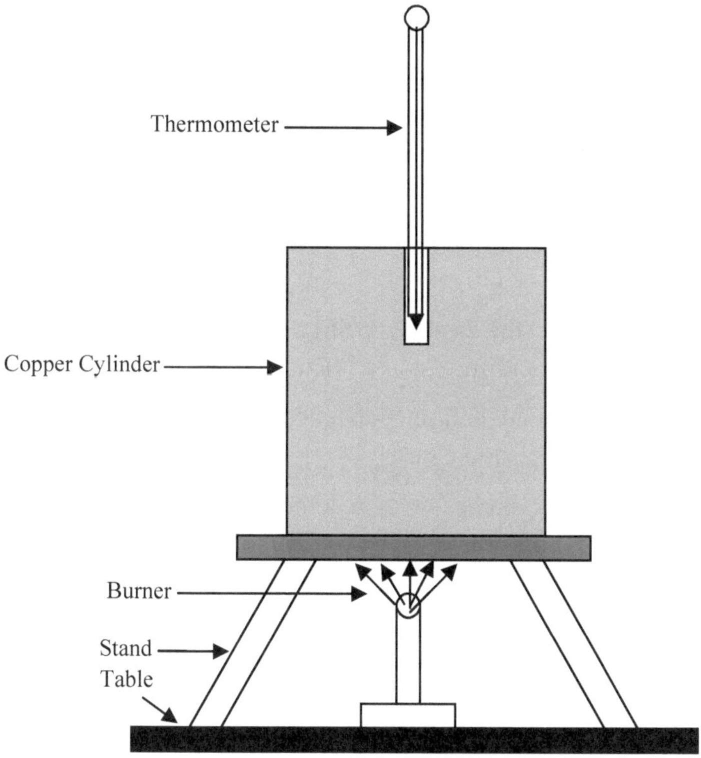

Fig. 3.1 Experimental setup for Determination of Radiation constant of Copper cylinder.

Bibliography :

1. Introduction to chemical engineering, 7th Edition, by **Walter L. Badger and Juliust T. Banchero** page No. 152 – 155.
2. Pharmaceutical Engineering principles and practices, 6th Edition, by **C.V.S. Subrahmanyam**., page No. 105 – 130.
3. Pharmaceutical Engineering, 2003 1st Edition, by **K. Sambamurthy,** page No. 75 – 77.
4. Cooper & Gun's, Tutorial Pharmacy, 6th Edition, Edited by **S.J. Carter**., Page No. 149 – 150.
5. Unit operations of chemical engineering, 6th Edition, by **M.C. Cabe, J.C.Smith, harriott,** page no. 425 – 426.

Experiment No : 04 **Dt :**

Heat Loss by Natural Convection and Radiation (Painted Flask)

Aim : To determine the heat loss through spherical body by natural convection and radiation from painted (Round bottom) flask.

Requirements : Round bottomed painted flask, beaker, thermometer, cork, stand with clamp, stop clock.

Principle : All solid bodies with a temperature above absolute zero, radiates energy. **Radiation** is a term given to the transfer of energy through space by means of *electromagnetic waves*. When heat flows by actual mixing of warmer portions with cooler portion of same material, the mechanism is known as **Convection**. Convection is restricted to the flow of heat in fluids. A black body radiates the maximum possible amount of energy at a given temperature. Stefan Boltzmann's law would give the total amount of radiation emitted by a block body, as follows;

$$q = \sigma A T^4$$

Where, q = Energy radiated per hour

A = Area of radiating surface

T = Absolute temperature of the radiating surface, °R

For black body the value of σ is 0.174×10^{-8} Btu/hr. sq. ft.°F^4.

No actual body radiates quite as much as the black body. The radiation by any actual body can be expressed as

$$q = \varepsilon \sigma A T^4$$

Where, ε = Emissivity of the body

Emissivity is a fraction less than 1 and the ratio of the energy emitted by the body that emitted by a black body at the same temperature.

For glass $\varepsilon = 0.92$; $C_p = 0.2$; $\sigma = 0.155$ in Btu/hr. sq. ft.°F^4.

Theory : Heat is transferred to the metal wall in four ways; by radiation from incandescent solids (fuel bed, brickwork, solid carbon in luminous flames), by hot gases, by conduction, and by convection. Any quantity of heat desired may be introduced per unit area of the metal surface, depends only on the temperature of the hot solid or gas and area of hot surface exposed to unit area of cold surface. This heat must pass through the viscous film and is therefore dependent on all the properties of the gas stream. The amount of thermal energy radiated by a surface is increased rapidly with increasing temperature, when heat flows by actual mixing of warmer portions with cooler portions of the same material. This mechanism is known as "**Convection**".

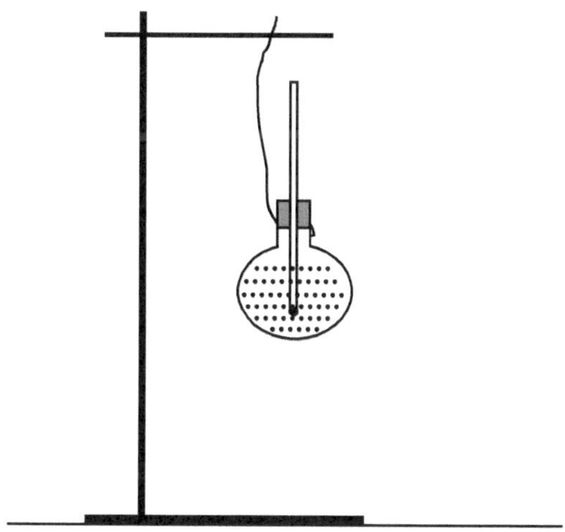

No paint on flask

Procedure :

1. Take a round bottom flask, measure the diameter, average radius, and then surface area is determined whose heat loss to be calculated.
2. The flask neck is covered with a black carbon paper, and is hanged in air by tying one end for neck with a thread, and other end to a clamp of stand.
3. Boil the water up to its boiling point and taken in to the flask up to the neck level.
4. The flask is fitted with a rubber cork having one hole, which is fitted with thermometer.
5. The temperature is noted for every 5 min till it reaches to room temperature.

6. A graph is plotted between **temperature** on Y–axis and **time** on X–axis.
7. By finding three different slopes, calculate the average and determine the rate of heat loss.

Precautions :

1. The hot water must be transferred as fast as possible.
2. Avoid parallax error while recording the temperature.

Observation :

S. No.	Time (Min)	Temp. (°C)
01	0	100
02	5	
03	10	
04	15	
05	20	
06	25	
07	30	
08	35	
09	40	
10	45	
11	50	
12	55	
13	60	
14	65	
15	70	
16	75	
17	80	

(M) Wt. of Round bottom flask = gm

Volume of water = ml

Determination of Area (A)

Volume of sphere = $4/3\pi r^3$

Area = $4\pi r^2$

$°F = 9/5\,°C + 32$ [For T_1 & T_2]

$^{o}R = {^{o}F} + 459.4$

$Q_C = M.\ C_p.\ d\theta/dt$

$[d\theta/dt = \text{average slope } (Y_2 - Y_1)/(X_2 - X1)]$

C_p = Specific heat of glass = 0.2

$Q_R = \sigma.A.\varepsilon\ (T_1^4 - T_2^4)$

(in **Btu/hr. Sq. ft. $^{o}F^4$**) for Glass – ε =0.92

$Q = Q_C + Q_R$ (in Btu/hr)

Specific Heat of Water = 0.95 – 0.963

Graphs : Determination of slopes

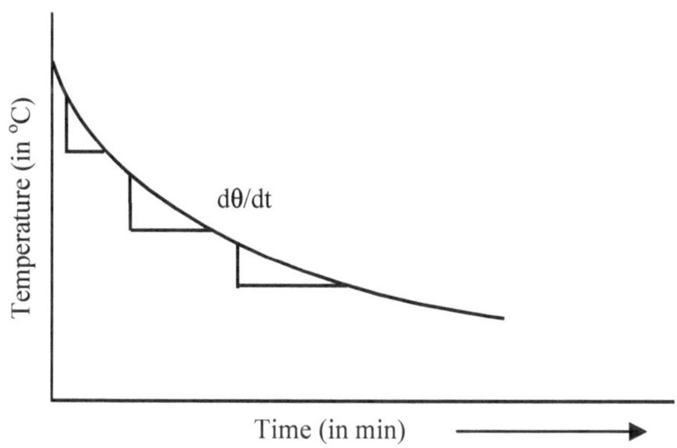

Report :

Bibliography :

1. Introduction to chemical engineering, 7th Edition, by **Walter L. Badger and Juliust T. Banchero** page No. 117 & 152 – 156.
2. Pharmaceutical Engineering principles and practices, 6th Edition, by **C.V.S. Subrahmanyam**., page No. 105 – 130.
3. Pharmaceutical Engineering, 2003 1st Edition, by **K. Sambamurthy,** page No. 75 – 77.
4. Cooper & Gun's, Tutorial Pharmacy, 6th Edition, Edited by **S.J. Carter**., Page No. 149 – 150.
5. Unit operations of chemical engineering, 6th Edition, by **M.C. Cabe, J.C.Smith, harriott,** page no. 425 – 426.

Experiment No : 05 **Dt :**

Heat Loss by Natural Convection and Radiation (Unpainted Flask)

Aim : To determine the heat loss through spherical body by natural convection and radiation from unpainted (Round bottom) flask.

Requirements : Round bottomed unpainted flask, beaker, thermometer, cork, stand with clamp, stop clock.

Principle : All solid bodies with a temperature above absolute zero, radiates energy. **Radiation** is a term given to the transfer of energy through space by means of *electromagnetic waves*. When heat flows by actual mixing of warmer portions with cooler portion of same material, the mechanism is known as **Convection**. Convection is restricted to the flow of heat in fluids. A black body radiates the maximum possible amount of energy at a given temperature. Stefan Boltzmann's law would give the total amount of radiation emitted by a block body, as follows;

$$\mathbf{q = \sigma A T^4}$$

Where, q = Energy radiated per hour

A = Area of radiating surface

T = Absolute temperature of the radiating surface, ^{o}R

For black body the value of σ is 0.174×10^{-8} Btu/hr. sq. ft.$^{o}F^4$.

No actual body radiates quite as much as the black body. The radiation by any actual body can be expressed as

$$\mathbf{q = \varepsilon \sigma A T^4}$$

Where, ε = Emissivity of the body

Emissivity is a fraction less than 1 and the ratio of the energy emitted by the body that emitted by a black body at the same temperature.

For glass $\varepsilon = 0.92$; $C_p = 0.2$; $\sigma = 0.155$ in Btu/hr. sq. ft.$^{o}F^4$.

Theory : Heat is transferred to the metal wall in four ways; by radiation from incandescent solids (fuel bed, brickwork, solid carbon in luminous flames), by hot gases, by conduction, and by convection. Any quantity of heat desired may be introduced per unit area of the metal surface, depends only on the temperature of the hot solid or gas and area of hot surface exposed to unit area of cold surface. This heat must pass through the viscous film and is therefore dependent on all the properties of the gas stream. The amount of thermal energy radiated by a surface is increased rapidly with increasing temperature, when heat flows by actual mixing of warmer portions with cooler portions of the same material. This mechanism is known as "**Convection**".

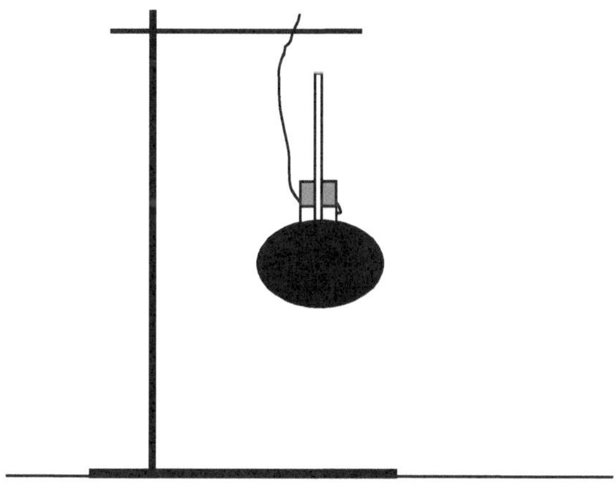

Painted on flask

Procedure :

1. Take a round bottom flask, measure the diameter, average radius, and then surface area is determined whose heat loss to be calculated.
2. The flask is hanged in air by tying one end for neck with a thread, and other end to a clamp of stand.
3. Boil the water up to its boiling point and taken in to the flask up to the neck level.
4. The flask is fitted with a rubber cork having one hole, which is fitted with thermometer.
5. The temperature is noted for every 5 min till it reaches to room temperature.
6. A graph is plotted between **temperature** on Y–axis and **time** on X–axis.
7. By finding three different slopes, calculate the average and determine the rate of heat loss.

Precautions :

1. The hot water must be transferred as fast as possible.
2. Avoid parallax error while recording the temperature.

Observation :

S. No.	Time (Min)	Temp. (°C)
01	0	100
02	5	
03	10	
04	15	
05	20	
06	25	
07	30	
08	35	
09	40	
10	45	
11	50	
12	55	
13	60	
14	65	
15	70	
16	75	
17	80	

(M) Wt. of Round bottom flask = gm

Volume of water = ml

Determination of Area (A)

Volume of sphere = $4/3\pi r^3$

Area = $4\pi r^2$

$^oF = 9/5\,^oC + 32$ [For T_1 & T_2]

$^oR = {}^oF + 459.4$

$Q_C = M.\ C_p.\ d\theta/dt$

[$d\theta/dt$ = average slope $(Y_2 - Y_1)/(X_2 - X1)$]

C_p = Specific heat of glass = 0.2

$Q_R = \sigma.A.\varepsilon\ (T_1^4 - T_2^4)$

(in **Btu/hr. Sq. ft. °F⁴**) for Glass– ε =0.92

$Q = Q_C + Q_R$ (in Btu/hr)

Graphs : Determination of slopes

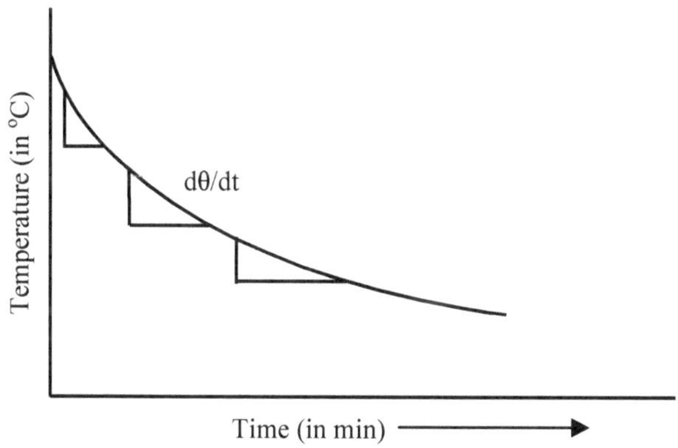

Report :

Bibliography :

1. Introduction to chemical engineering, 7th Edition, by **Walter L. Badger and Juliust T. Banchero** page No. 117 & 152 – 156.
2. Pharmaceutical Engineering principles and practices, 6th Edition, by **C.V.S. Subrahmanyam**., page No. 105 – 130.
3. Pharmaceutical Engineering, 2003 1st Edition, by **K. Sambamurthy,** page No. 75 – 77.
4. Cooper & Gun's, Tutorial Pharmacy, 6th Edition, Edited by **S.J. Carter**., Page No. 149 – 150.
5. Unit operations of chemical engineering, 6th Edition, by **M.C. Cabe, J.C.Smith, harriott,** page no. 425 – 426.

Experiment No : 06 **Dt :**

Size Reduction and Separation

Aim : To compare the efficiency of size reduction of a mixer and a ball mill.

Requirements : Ball mill, Mixer/Blender, Dill fruits, Coriander fruits, sieves # 10, 22, 44, 80, 100 & 120 and sieve shaker.

Objectives : The objectives of the size reduction are to get improved dissolution rate, improved rate of absorption, effective extraction of drugs, effective drying, improved physical stability, uniform flow and content uniformity, etc.

Theory : Size reduction is the process of reducing substances to small particles. This unit operation is also commonly known as *comminution* or *diminution* or *grinding* or *pulverization* and is extensively used in pharmaceutical operations both in the laboratory and on the industrial scale. Size reduction increases surface area of the material and the greater the surface area of the material the higher the rate of dissolution of the material in a suitable solvent. Normally size reduction may be achieved by two methods, namely ***precipitation*** or ***mechanical*** process.

There are four types of modes of stress applied in size reduction,

(i)	Cutting	Ex : Cutter mill
(ii)	Compression	Ex : Roller mill
(iii)	Impact	Ex : Hammer mill
(iv)	Attrition	Ex : Fluid energy mill

The ball mill consists of a horizontally rotating hallow vessel of cylindrical shape with the length slightly greater than its diameter. The mill is partially filled with balls of steel or pebbles. The load of balls in a ball mill is normally such that when the mill is stopped, the balls occupy about one–half the volume of the mill. The void fraction in the mass of balls, when at rest, is typically 0.40. The grinding may be done with dry solids, but more commonly the feed is a suspension of the particles in water, increasing both the capacity and the efficiency of the mill.

Principle : The ball mill works on the principle of ***Impact*** and ***Attrition*** between the rapidly moving balls and the powder material, both enclosed in a hallow cylinder. At low speeds, the balls roll over each other and attrition (rubbing action) will be the predominant mode of action. Thus in the ball mill, impact or attrition or both are responsible for the size reduction.

In the cutter mill, size reduction involves successive cutting or shearing the feed material with the help of sharp knives or blades. This is used to obtain a coarse degree of size reduction of soft materials. The commonest application is the treatment of drugs such as roots, peels or woods, and prior to extraction.

Procedure :

1. Take equally weighed (5 gm. each) amounts of *CORIANDER* and *DILL* fruits.
2. Divide into two successive portions (each of 2.5 gm) in ball mill and Mixer/Blender respectively.
3. Both are operated for 15 & 30 min. The powdered drug is put in a set of sieves (#10, 22, 44, 80, 100 & 120) and undergone for shaking with the help of sieve shaker for about 10 min.
4. The amount retained on each mesh is noted by weighing of individual powder.
5. The same process is repeated for Dill fruits and amount retained on each sieve is noted.
6. Finally find out the percentage amount retained on each sieve and give report.

Observation : After 15 min

Table 6.1 Coriander obtained from MIXER/BLENDER.

S. No.	Sieve No.	Amt. Retained (in gms)	Percentage retained
1	10		
2	22		
3	44		
4	80		
5	100		
6	120		

Table 6.2 Coriander obtained from BALL MILL.

S. No.	Sieve No.	Amt. Retained (in gms)	Percentage retained
1	10		
2	22		
3	44		
4	80		
5	100		
6	120		

Table 6.3 Dill obtained from MIXER/BLENDER.

S. No.	Sieve No.	Amt. Retained (in gms)	Percentage retained
1	10		
2	22		
3	44		
4	80		
5	100		
6	120		

Table 6.4 Dill obtained from BALL MILL.

S. No.	Sieve No.	Amt. Retained (in gms)	Percentage retained
1	10		
2	22		
3	44		
4	80		
5	100		
6	120		

After 30 min

Table 6.5 Coriander obtained from MIXER/BLENDER.

S. No.	Sieve No.	Amt. Retained (in gms)	Percentage retained
1	10		
2	22		
3	44		
4	80		
5	100		
6	120		

Table 6.6 Coriander obtained from BALL MILL.

S. No.	Sieve No.	Amt. Retained (in gms)	Percentage retained
1	10		
2	22		
3	44		
4	80		
5	100		
6	120		

Table 6.7 Dill obtained from MIXER/BLENDER.

S. No.	Sieve No.	Amt. Retained (in gms)	Percentage retained
1	10		
2	22		
3	44		
4	80		
5	100		
6	120		

Table 6.8 Dill obtained from BALL MILL.

S. No.	Sieve No.	Amt. Retained (in gms)	Percentage retained
1	10		
2	22		
3	44		
4	80		
5	100		
6	120		

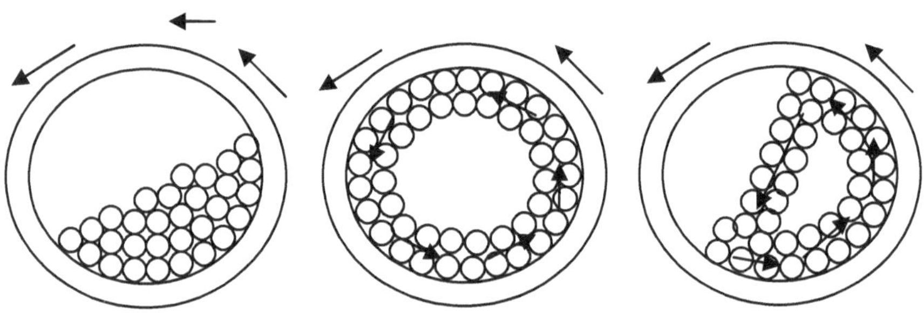

A
Low Speed

B
High Speed

C
Correct Speed

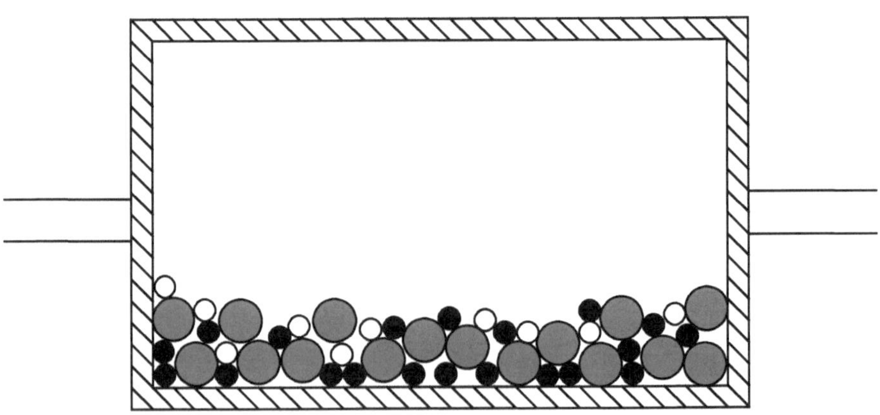

Fig. 6.1 Side view of Experimental setup of Ball mill.

Report :

Bibliography :

1. Introduction to chemical engineering, 7th Edition, by **Walter L. Badger and Juliust T. Banchero** page No. 677 – 678.
2. Pharmaceutical Engineering principles and practices, 6th Edition, by **C.V.S. Subrahmanyam**., page No. 147 – 158.
3. Cooper & Gun's, Tutorial Pharmacy, 6th Edition, Edited by **S.J. Carter**., Page No. 185 – 189.
4. Pharmaceutical Engineering, 2003 1st Edition, by **K. Sambamurthy** page No. 339 – 342.
5. Unit operations of chemical engineering, 6th Edition, by **M.C. Cabe, Smith, harriott,** page no. 976 – 977.
6. The theory and practice of Industrial Pharmacy, 3rd Edition, by **Leon Lachman, Herbert A. Lieberman,** page no: 40 – 41.

Experiment No : 07 **Dt:**

Flow of Fluids
(Frictional Factor)

Aim : To determine the Reynold's Number and frictional factor by calculating frictional loses in pipes in which fluids are flowing.

Requirements : Tubes of pipes

Measuring cylinders (1000 ml)

Vernier calipers, and beakers

Objectives :

(i) Passing of reactants (liquids or gases) into the reaction system.

(ii) Transferring of air, nutrient broth into the fermenter.

(iii) Bottling of liquid (dosage forms) medicaments into suitable containers.

(iv) Transportation of sterile air and sterile water in the production of parenterals.

(v) Mixing of solids and liquids in case of suspensions.

(vi) Packing of semisolids in containers.

Principle : The term '***fluids***' include both liquids and gases. The flow of fluids through a closed channel is influenced by various factors. The flow can be either viscous or turbulent and this can be observed in classical Reynold's experiment. When a fluid is passing through a pipe of uniform diameter at very slow motion, one can observe a laminar flow. This is because, the fluid flow is parallel straight lines and this is also known as *viscous* or *stream line flow*.

There are two forms of fluid motions, ***viscous flow*** and ***turbulent flow***. The velocity at which the flow changes from one kind of flow to other is known as the ***critical velocity***. The critical velocity depends on diameter of tube, velocity of fluid, density of liquid, Viscosity of liquid. By knowing Reynold's Number we can determine the type of flow, the liquid exhibit.

The factors, which affect the mechanism of fluid flow, are diameter–D, average velocity 'u', density of liquid 'ρ', and viscosity of the fluid 'μ'.

Reynold's Number (R_e) can be calculated as

$$R_e = Du\rho / \mu$$

Where, D – Diameter of pipe/tube (m)

u – Velocity of fluid (m/s)

ρ – density of fluid (for water = 1 kg/m^3)

μ – viscosity of fluid (for water = 0.01 Poise or kg/m.s.)

Reynold's number is a dimensionless group. For straight circular pipes, the value of R_e is less than 2,100 and flow will always be viscous/stream line. When value of R_e is over 4,000, the flow will always be turbulent. Between 2,100 and 4,000 a *Transition region* is found where the flow may be either laminar or turbulent, depending upon conditions at the entrance of the tube and on the distance from the entrance.

Frictional Factor : In the fluid flowing through a circular pipe of length, the total force resisting the flow must equal the product of the area of contact between the fluid and pipe wall. The pressure drop will equal this product divided by the cross-sectional area of the pipe, since pressure is measured in force per unit area. This can be represented in terms of Reynold's number and frictional factor.

$$f = 16 / R_e$$

Procedure :

1. Take a jar, and fill the jar with one liter of water, and mark the upper level of the water with the permanent marker.
2. Now, select different types of pipes, and measure their diameter.
3. Then connect one end of the pipes to reservoir of water such that, the velocity of water flowing through tube can be varied. Three different pipes of varying diameter are connected to three different taps.
4. From each pipe collect water (1 lt.) each time by varying speeds (Low, Medium & High). Time taken for the collection of water in each case is noted.
5. Calculate the velocity of water.

6. From the readings taken, R_e (Reynold's number) and frictional factor are calculated.
7. Plot a graph between R_e (Reynold's number) and frictional factor (f).

Report : From the observation made, it is inferred that, as velocity of flow of fluid increases, Reynold's number increases while the frictional factor decreases. With increase in diameter of the pipe, Reynold's number increases.

Observation : Reynold's Number

$$R_e = Du\rho / \mu$$

Frictional factor can be calculated by using formula—

$$f = 16 / R_e$$

Table 7.1

Pipes	Time taken to collect 1000 ml water	Diameter of pipe (D)	Radius of pipe (r)	Area A (πr^2)	Avg. velocity = volume/ Area X time	R_e = $Du\rho/\mu$	Frictional factor (f) = $16/R_e$

Graph :

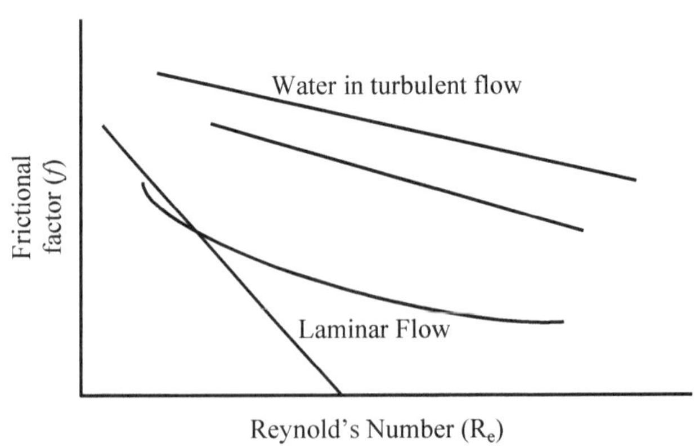

Graph : Graph between R_e (Reynold's number) and frictional factor (*f*).

Bibliography :

1. Introduction to chemical engineering, 7th Edition, by **Walter L. Badger and Juliust T. Banchero** page No. 31 – 44.
2. Cooper & Gun's, Tutorial Pharmacy, 6th Edition, Edited by **S.J. Carter**., Page No. 146- 148.
3. Pharmaceutical Engineering principles and practices, 6th Edition, by **C.V.S. Subrahmanyam**., page No. 25 – 35.
4. Pharmaceutical Engineering, 20031st Edition, by **K. Sambamurthy** page No. 16 – 18.
5. Unit operations of chemical engineering, 6th Edition, by **M.C. Cabe, Smith, harriott,** page no. 113 – 115.

Experiment No : 08 **Dt :**

Flow of Fluids
(Reynold's Number)

Aim : To determine the Reynold's Number by calculating Velocity of fluids & Area of pipes in which fluids are flowing.

Requirements : Tubes of pipes

Measuring cylinders (1000 ml)

Vernier calipers, and beakers

Objectives :

(i) Passing of reactants (liquids or gases) into the reaction system.

(ii) Transferring of air, nutrient broth into the fermenter.

(iii) Bottling of liquid (dosage forms) medicaments into suitable containers.

(iv) Transportation of sterile air and sterile water in the production of parenterals.

(v) Mixing of solids and liquids in case of suspensions.

(vi) Packing of semisolids in containers.

Principle : The term '***fluids***' include both liquids and gases. The flow of fluids through a closed channel is influenced by various factors. The flow can be either viscous or turbulent and this can be observed in classical Reynold's experiment. When a fluid is passing through a pipe of uniform diameter at very slow motion, one can observe a laminar flow. This is because, the fluid flow is parallel straight lines and this is also known as viscous or stream line flow.

There are two forms of fluid motions, ***viscous flow*** and ***turbulent flow***. The velocity at which the flow changes from one kind of flow to other is known as the ***critical velocity***. The critical velocity depends on diameter of tube, velocity of fluid, density of liquid, Viscosity of liquid. By knowing Reynold's Number we can determine the type of flow, the liquid exhibit.

The factors, which affect the mechanism of fluid flow, are diameter–D, average velocity 'u', density of liquid 'ρ', and viscosity of the fluid 'μ'.

Reynold's Number (R_e) can be calculated as

$$\mathbf{R_e = Du\rho / \mu}$$

Where, D – Diameter of pipe/tube (m)

u – Velocity of fluid (m/s)

ρ – density of fluid (for water = 1 kg/m^3)

μ – viscosity of fluid (for water = 0.01 Poise or kg/m.s.)

Reynold's number is a dimensionless group. For straight circular pipes, the value of N_{RE} is less than 2,100 and flow will always be viscous/stream line. When value of N_{RE} is over 4,000, the flow will always be turbulent. Between 2,100 and 4,000 a *Transition region* is found where the flow ma be either laminar or turbulent, depending upon conditions at the entrance of the tube and on the distance from the entrance.

Significance of Reynold's Number :

- It can be used to predict the nature (viscous or turbulent) of flow in a particular set of circumstances.
- The physical stability of suspension depends on the rate of settling of the particles.
- Rate of sedimentation of particles must not be too rapid to create turbulence. Stoke's Law is modified to include Reynold's number. Therefore, the type of flow (whether laminar or turbulent) is important.
- The rate of heat transfer in liquids also depends on the flow, whether viscous or turbulent, etc.

Procedure :

1. Collect the two water reservoirs, one for colour water and another one, normal water.
2. Get a pipe line/tube line, which is transparent and should be connected one end to the coloured water reservoir, and normal water reservoir connected in between pipe line as showed in diagram.
3. Arrange (connect) the experimental setup as shown in diagram below.

4. In Reynold's experiment, a glass tube is connected to a reservoir of water in such a way that the capacity of water flowing through the tube could be varied at will.
5. In the inlet end of the tube a nozzle is inserted through which a fine stream of coloured water could be introduced.

Conclusion : Reynold's found that when the velocity of the water was low, the thread of colour maintained itself throughout the tube. By putting in more than one of these jets at different points in cross section, it can be shown that in no part of the tube was there any mixing, and the fluid flowed in parallel, straight lines.

Observation : Reynold's Number

$$R_e = Du\rho / \mu$$

Table 8.1

Pipes	Time taken to collect 1000 ml water	Diameter of pipe (D)	Radius of pipe (r)	Area A (πr^2)	Avg. velocity = volume/Area X time	R_e = $Du\rho/\mu$

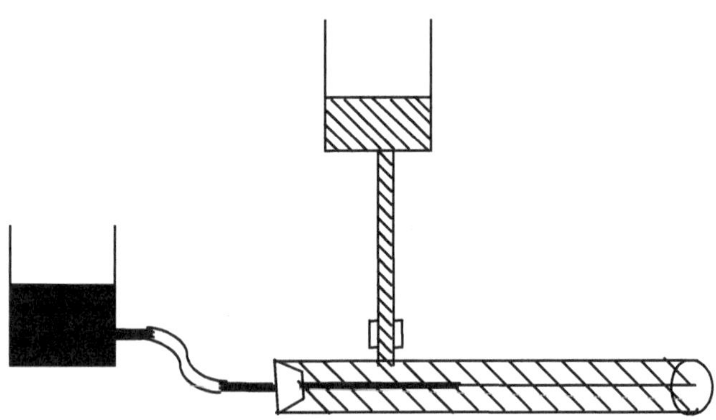

Fig. 8.1 Reynold's Experimental Setup.

Report :

Bibliography :

1. Introduction to chemical engineering, 7th Edition, by **Walter L. Badger and Juliust T. Banchero** page No. 31 – 44.
2. Cooper & Gun's, Tutorial Pharmacy, 6th Edition, Edited by **S.J. Carter**., Page No. 146- 148.
3. Pharmaceutical Engineering principles and practices, 6th Edition, by **C.V.S. Subrahmanyam**., page No. 25 – 35.
4. Pharmaceutical Engineering, 2003 1st Edition, by **K. Sambamurthy** page No. 16 – 18.
5. Unit operations of chemical engineering, 6th Edition, by **M.C. Cabe, Smith, harriott,** page no. 51 – 53.

Experiment No : 09 **Dt :**

Rate of Drying

Aim : To determine the rate of drying for the given sample.

Requirement : Petri dish, Hot air oven, Thermometer, Spatula, $CaCO_3$, Weighing balance, and weights

Principle : Drying is a ***unit operation***. It is a process of removal of trace amount of elements (moisture) from a liquid (or) gas (or) solid (or) semi-solid, at temperature lower than boiling point of materials.

Heat and mass transfers are involved in this process. Drying is a process, which ensures the stability of the product. Tray dryer is a static bed dryer and mechanism of heat transfer is "**Convection**". It is used for drying of crude drugs, granules etc.

Theory : The experimental data obtained in an investigation of the effect of external conditions on the drying of a solid by air are usually the moisture content of the solid as a function of time under *constant drying conditions*. The term constant drying conditions means that the air velocity, temperature, humidity, and pressure are maintained constant and that the outlet air conditions are substantially the same as those at the inlet.

The behavior of drying of solids is explained by drying curve. The time required for drying is a batch weight of material in dryer can be estimated with the help of "***Drying Curve***". The drying–rate curve may be divided into a *constant rate period*, such as the portion 'AB' in the following diagram, and the *falling–rate period* 'BD'. The free moisture content at point 'B' is called the *critical moisture content*. If the desired moisture content is larger than the critical moisture content, only the constant–rate period will occur.

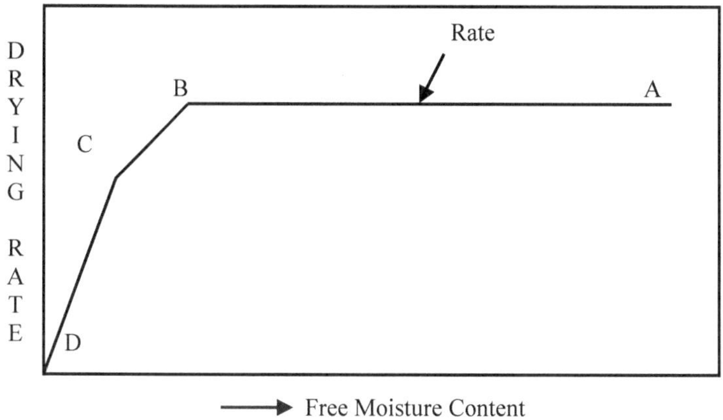

The rate of drying can be calculated as the ratio of the moisture lost per Sq. cm. per given time. The rate of drying is a standard parameter for several drugs, and can be determined by calculating the amount of water, it need to take (or) to makeup slurry and calculated by a series of drying under consecutive time intervals. The rate of drying can be calculated by the formula.

$$\textbf{Rate Of Drying} = (W_n - W_{n+1}) / \Delta t.A$$

Where, A = Area of the plate expose to drying

W_n = Weight of glass dish + sample slurry

W_{n+1} = Weight of glass dish + sample slurry after time "t".

Determine the diameter of the given glass dish and calculate the area ($A = \pi r^2$), Δt–time interval.

Procedure :

1. Take a clean Petri dish without lid and determine its weight and area of the petri dish, as "W_1", "A" respectively.
2. Take the given powder sample (10 gm. of $CaCO_3$) and fill the petri dish & determine its weight 'W_2'.
3. Make the powder into slurry in the petri dish with the water, and add little more in it, so that a thin layer of water forms above. Let its weight be 'W_3' gm.
4. Now, the petri dish containing slurry is kept in drier (Hot Air Oven) at 70 °C.
5. Continue drying & determine the weight of the sample at every 15 min. interval of time.

6. Stop drying process, when there is no change in the weight of the petri dish (Sample slurry + Petri dish).
7. Determine average moisture content & drying rate for each time interval.
8. Plot a graph between "**Rate of dying** on Y – axis and **Avg. moisture content** on X – axis".
9. Plot a graph between "**Rate of dying** on Y – axis and **Time interval** on X – axis".

Observation : Area of petri dish ($A = \pi r^2$) = Sq.cm.

Wt. of empty petri dish (W_1) = gm

Wt. of empty petri dish + Sample (W_2) = gm

Wt. of empty petri dish + Sample + Water (W_3 or W_n) = gm

Wt. of Sample ($CaCO_3$) = 10 gm

Wt. of Water added ($W_3 - W_2$) = gm

Wt. of the sample after 15 min. drying (W_4 or W_{n+1}) = gm

Moisture Content (MC) = $(W_3 – W_2) / (W_2 – W_1)$ = gm

At '0' time, MC_1 =

After '15 min' time, MC_2 =

Avg. Moisture Content $[(MC_1 + MC_2) / 2]$ = gm

Rate of drying = $(W_n - W_{n+1})/\Delta t.A$ = gm

Graphs :

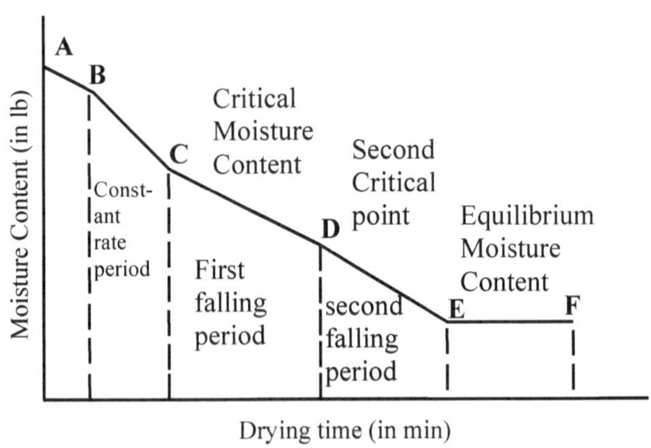

Graph 1 : The graph between **Moisture content** Vs **Avg. Moisture content**

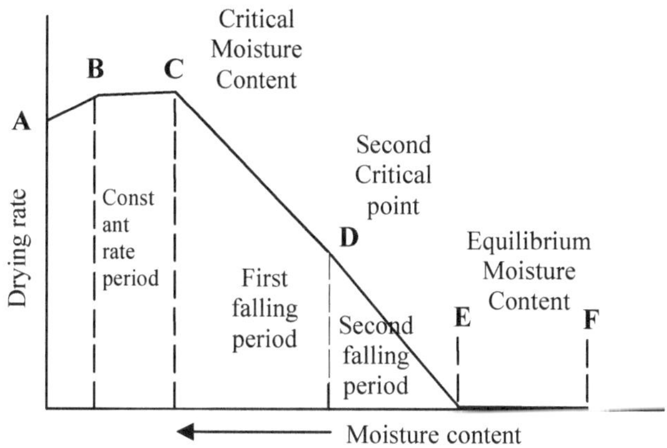

Graph 2 : The graph between **Rate of drying** Vs **Avg. Moisture content**

Petri dish :

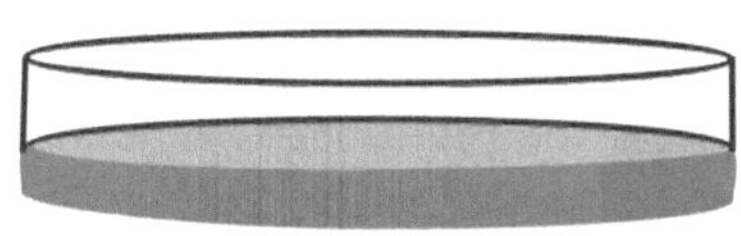

Report :

Bibliography :

1. Introduction to chemical engineering, 7th Edition, by **Walter L. Badger and Juliust T. Banchero** page No. 491 – 501.
2. Pharmaceutical Engineering principles and practices, 6th Edition, by **C.V.S. Subrahmanyam**, page No. 382 389.
3. Cooper & Gun's, Tutorial Pharmacy, 6th Edition, Edited by **S.J. Carter**., Page No. 273 – 275.
4. Pharmaceutical Engineering, 2003 1st Edition, by **K. Sambamurthy** page No. 318 – 322
5. Unit operations of chemical engineering, 6th Edition, by **M.C. Cabe, Smith, harriott,** page no. 782 – 788.

Experiment No : 10 **Dt :**

Particle Size Analysis

Aim : To determine the average particle size of each fraction by sieving (sieve analysis) & to construct frequency size distribution plot & cumulative size distribution curves.

Requirement : Sieve No.s # 10, 20, 44, 80, 100 and 120

Powder samples (NaCl & $NaHCO_3$)

Principle : Sieve analysis is the size separation method used to determine the average particle size of a given powder in the laboratories, different grades of sieves are used for sifting the powder & for preparing granules of required size. After comminuting any material is passed through a specific sieve grade to get the required particle size, bigger particles are retained in the machine or apparatus for regrinding, after final grinding & sieving the material, which does not pass through the sieve, is known as "Tailing".

Sieving is most widely used method for measuring particle size distribution because of inexpensive, simple & rapid with little variation between operators.

Sieve Shaker Machine : The powdered drug is separated according to its particle size, using number of sieves in a nest. These are subjected to different types of agitation, so that the size separation is rapid.

Theory : Screening is a method of separating particles according to size alone. In industrial screening the solids are dropped on, or thrown against, a screening surface. The undersize particles, or *fines*, pass through the screen openings; oversize particles, or *tails*, do not. Solids of small size are generally specified according to their screen analysis. A screen analysis of a material is carried out by placing a sample on the coarsest of a set of standard screens. Below this screen are arranged the remaining screens in the series in order of decreasing size of mesh. The pile of screens with the sample on the top screen is shaken in a definite manner, either manually or mechanically, for a definite length of time, and the material collected on each screen is removed and weighed. Screen analysis may show the weight percentage of the material that passes through 10 mesh and remains on 20 mesh, through 20 and 30, through 30 and 40, etc. such a report is quite meaningless and should never be employed unless the screen themselves are specified.

Procedure :

1. Select, sets of standard sieves (or) screens & arrange that, the finest sieve be at the bottom with other sieves placed in the order of increasing size opening (i.e. in the order of decreasing sieve number).

   Place a receiver below the bottom screen & cover on the top screen.
2. Weigh accurately 50 gms of a given powder sample & place on the top screen (coarse). The cover is placed & the pile of screens is shaken in a definite manner, either manually or mechanically for a definite length of time (20 min.).
3. Weigh the powder fractions retained on each sieve and determine the rate of percentage of each powder sample retained on every sieve.
4. Calculate the % wt. retained on sieve using the following formula.

$$\frac{\text{Wt. of the powder retained on sieve}}{\text{Total wt. of powder taken for experiment}} \times 100$$

5. Plot a graph of % wt. retained on Y–axis against log avg. particle size (in microns) on X–axis and cumulative % wt. retained on Y–axis against log screen opening (in microns) on X–axis.

Graphs :

1. Frequency distribution curve / plot

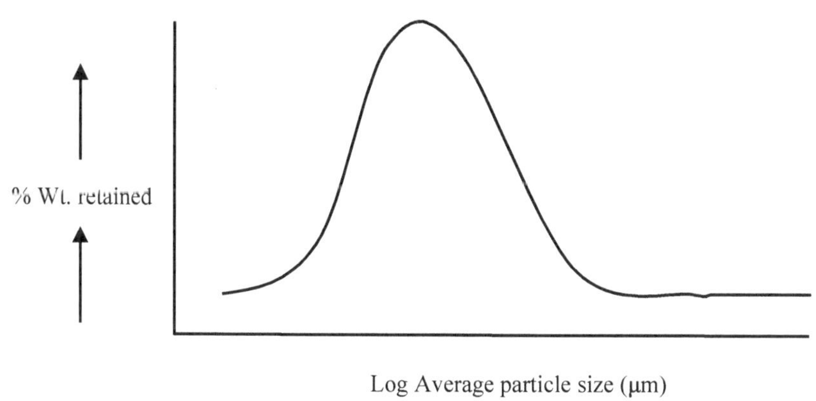

2. Cumulative size distribution curve / plot

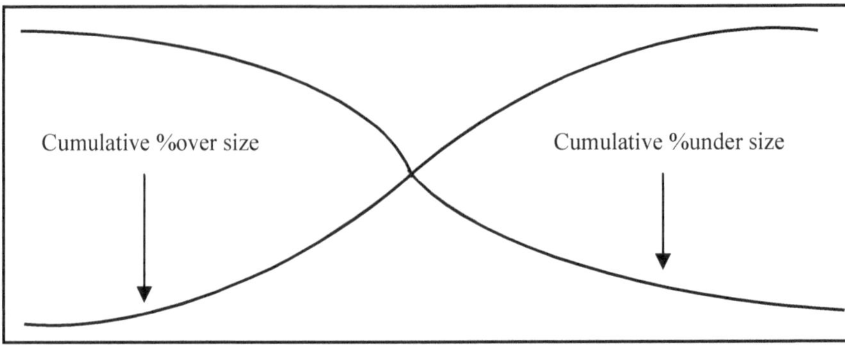

Log Screen opening (μm)

Observation :

Sl. No.	Sieve No.	Screen Opening	Avg. Particle Size	% Wt. retained	Cumulative % of Over Size	Cumulative % of Under Size
1	10	1651	-			
2	22	710	1180.5			
3	44	355	761.75			
4	80	175	473.87			
5	100	150	311.93			
6	120	125	218.46			

Example: Procedure for calculation of Cumulative % oversize & undersize.

Report :

Bibliography :

1. Introduction to chemical engineering, 7th Edition, by **Walter L. Badger and Juliust T. Banchero** page No. 618 – 622.
2. Pharmaceutical Engineering, 2003 1st Edition, by **K. Sambamurthy** page No. 355 – 359 & 375 – 377.
3. Pharmaceutical Engineering principles and practices, 6th Edition, by **C.V.S. Subrahmanyam**, page No. 183 – 189.
4. Cooper & Gun's, Tutorial Pharmacy, 6th Edition, Edited by **S.J. Carter**., Page No. 185 – 189.
5. Unit operations of chemical engineering, 6th Edition, by **M.C. Cabe, Smith, harriott,** page no. 986 – 991.
6. The theory and practice of Industrial Pharmacy, 3rd Edition, by **Leon Lachman, Herbert A. Lieberman,** page no: 27.

Experiment No : 11 **Dt :**

Effect of Filter Aids on Rate of Filtration

Aim : To study the effect of Filter Aids on the rate of filtration and to determine the optimum concentration of filter aid.

Requirement : Filter Aids (Bentonite & Talc), Calcium carbonate, Water

Principle : The filter aid is a finely divided solid material, but consisting of hard strong particles that are *masse* incompressible, for sludge's that are difficult to filter. Filter aid forms a surface deposit, which screens out the solids and also for events the plugging of the filter media. The various filter aids are used in the pharmaceutical field, depending on their usage and availability.

Ex : Kieselguhr/diatomaceous earth, asbestos, talc and charcoal etc.

The most important filter aid, widely used is diatomaceous earth or kieselguhr. It consists of siliceous skeletons of very small marine organisms known as 'diatoms.'

Theory : Slimy or very fine solids that form a dense impermeable cake quickly plug any filter medium that is fine enough to retain them. Practical filtration of such materials requires that the porosity of the cake be increased to permit passage of the liquor at a reasonable rate, this is done by adding a filter aid.

Filter aid can be used in either of *three* (3) ways.

1. Use of *pre-coat* of filter aid or thin layer of the material lay down on the filter before the sludge proper is pumped to the apparatus. A pre-coat prevents the colloidal particles of the sludge from becoming so enlarged in the filter cloth that, the resistance of the end of filtration.
2. Using filter aid is the incorporation of a certain percentage of the material with the sludge before sending it to the press. The presence of the filter aid increases the porosity of the sludge, decease its compressibility, and reduces the resistance of the cake during filtration.

3. Using filter aid is the use of special *pre-coat filter*. This is essentially a rotary–drum vacuum filter. Slurry of filter aid only is fed to the filter until a layer of pre-coat 2″ (inch) or more thick has been laid down. Then sludge to be filtered is fed. The doctor knife is so positioned that it peels off the sludge and an extremely thin layer of the pre-coat. Filtration is then interrupted, a new thick layer of pre-coat deposited, and filtration continued. This method is used for slimy or gelatinous precipitates that can never be built up to a cake but must be removed when the filter has on it a very thin layer of the precipitate.

The object of the filter aid is to prevent the medium from becoming blocked and to form an open porous cake. Hence, reduce the resistance to flow of filtering solution.

Procedure :

1. Weigh accurately 5 gm of calcium carbonate, transfer to graduated conical flask and add water to dissolve Calcium carbonate and make up volume up to 100 ml.
2. Prepare solutions like step No: 1 for about 6 numbers in different conical flasks.
3. Then add bentonite in the concentrations of 0.1 %, 0.2 %, 0.3%, 0.4% and 0.5% to the above five (5) solutions.
4. Number the conical flasks with numerical numbers like 1, 2, 3, 4, 5 & 6.
5. Now, pass the first sample (Control) [i.e., 0.0% bentonite in 5% calcium carbonate] marked as No: 1 through a selected filter paper and note down time taken for filtration.
6. Similarly, repeat the entire experiment for the other remaining solutions and determine time required for filtration and record them in the observation by using the following table.

Determination of Optimum Concentration :

To know the optimum concentration of filtration aid, draw a graph by taking filter aid on X-axis and rate filtration on Y- axis.

As the concentration of filter aid is increased, rate of filtration also increases but the point reaches where further additions of filter aid decreases the rate of filtration. The concentration at this change is optimum concentration of filter aid.

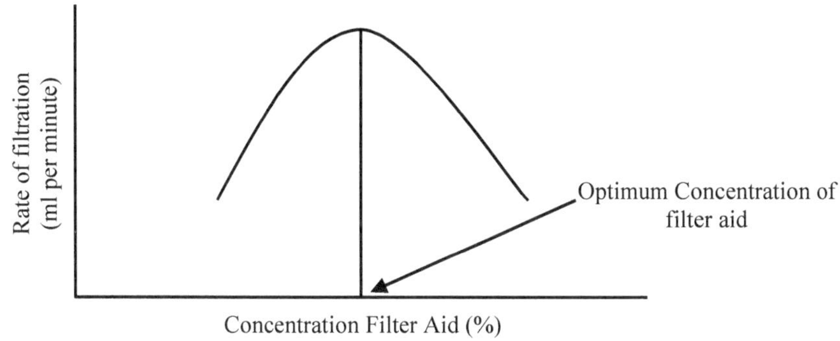

Observations :

S. No	Volume of 5% Calcium carbonate	Concentration of filter aid (%)	Time taken for filtration (minutes)	Rate of filtration (ml per minute)
1	100 ml	0.0		
2	100 ml	0.1		
3	100 ml	0.2		
4	100 ml	0.3		
5	100 ml	0.4		
6	100 ml	0.5		

Report :

Bibliography :

1. Pharmaceutical Engineering, 2003 1st Edition, by **K. Sambamurthy,** page No. 248 – 250.
2. Introduction to chemical engineering, 7th Edition, by **Walter L. Badger and Juliust T. Banchero** page No. 573 – 574.
3. Cooper & Gun's, Tutorial Pharmacy, 6th Edition, Edited by **S.J. Carter**., Page No. 185 – 189.
4. Pharmaceutical Engineering principles and practices, 6th Edition, by **C.V.S. Subrahmanyam**, page No. 183 – 189.
5. Unit operations of chemical engineering, 6th Edition, by **M.C. Cabe, Smith, harriott,** page no. 1003 – 1004.
6. The theory and practice of Industrial Pharmacy, 3rd Edition, by **Leon Lachman, Herbert A. Lieberman,** page no: 149 – 152.

Experiment No : 12 **Dt :**

Factors Affecting the Rate of Filtration

Aim : To study the effect of various factors on the rate of filtration.

Apparatus : Filter paper, vacuum pump, buchner funnel, stop clock

Principle : Filtration is a process whereby solid particles present in a suspension are separated from the liquid or gas employing a porous medium, which retains the solid but allows the fluid to pass through. The volume of the filtrate obtained through the filter paper per unit time is called Rate of filtration. The rate of filtration can be given by mathematical equation.

$$dv/dt = K.A.\Delta P/ \mu.l.$$ → **Darcy's Law**

Where A = Area of filter

ΔP = Pressure drop across the filter medium & cake

μ = Viscosity of filtrate

l = thickness of cake

V = Volume of the filtrate

t = time taken for filtration

K = Constant for the filter medium & filter cake (or) Resistance

Procedure :

Effect of Thickness of Cake : Prepare two solutions of calcium carbonate using water as the solvent; the concentrations of the solutions are 5% & 10 % respectively. Filter them and note the time taken for filtration to calculate the rate of filtration & compare them.

Effect of Viscocity : (using two solutions, one with water & other with mixture of glycerin and water (20:80 ratio) respectively).

Prepare two different solutions of 5% $CaCO_3$ using above prepared water & glycerin mixture. Filter them and note the time for filtration to calculate the rate of filtration & compare them.

Effect of Area : This can be determined by using funnel of large (big) and small areas for the same concentration of the solution (5% $CaCO_3$). Times taken for filtrations are noted, & calculate the rate of filtration and compare them.

Effect of Pressure : Prepare two solutions of calcium carbonate (5% of each) using water as solvent. Filter one of the solutions through a Buchner funnel, which is connected to a suction pump, and the other one is filtered through without suction pump and note the time taken for filtration, calculate the rate of filtration & compare them.

Observation :

1. **Effect of Thickness :**

Sample : 50 ml of 5% $CaCO_3$

50 ml of 10% $CaCO_3$

Sample	Time taken for filtration	Vol. of filtrate	Rate of filtration
5% $CaCO_3$			
10% $CaCO_3$			

2. **Effect of Viscosity :**

Sample : 50 ml of 5% $CaCO_3$

50 ml of 5% $CaCO_3$ [glycerin water mixture (20: 80 :: glycerin : water)]

Sample	Time taken for filtration	Vol. of filtrate	Rate of filtration
5% $CaCO_3$			
5% $CaCO_3$ with glycerin water mixture			

3. Effect of Area of the Funnel :

Sample : 50 ml of 5% $CaCO_3$

Sample	Time taken for filtration	Vol. of filtrate	Rate of filtration
5% $CaCO_3$ filter in small funnel			
5% $CaCO_3$ filter in big funnel			

4. Effect of Pressure :

Sample : 50 ml of 5% $CaCO_3$ [glycerin water mixture (20: 80 :: glycerin : water)]

Sample	Time taken for filtration	Vol. of filtrate	Rate of filtration
5% $CaCO_3$ filter with Pressure			
5% $CaCO_3$ filter without Pressure			

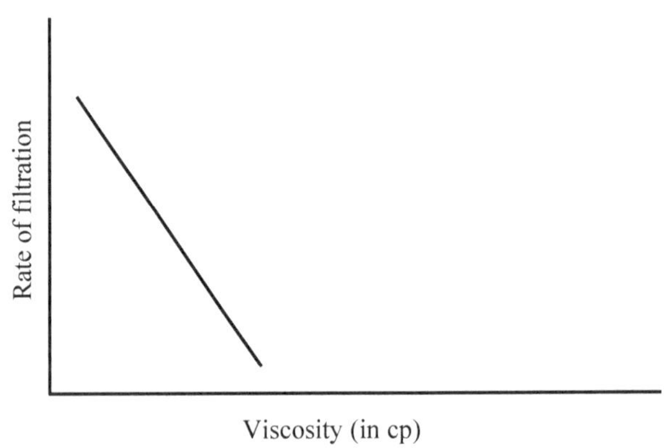

A graph, between **Rate of filtration Vs Viscosity**

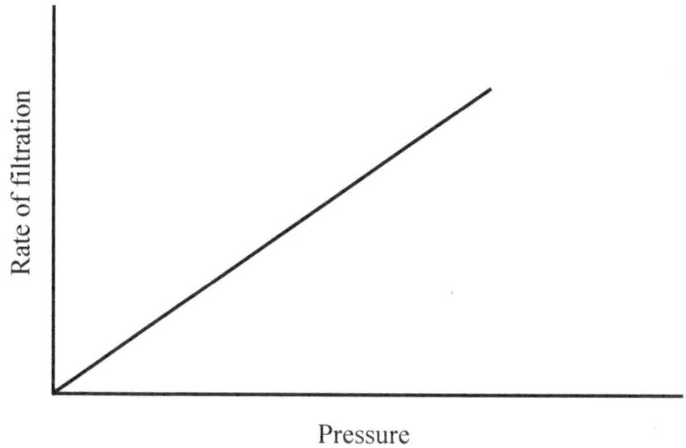

A graph, between **Rate of filtration Vs Pressure**

Report :

Bibliography :

1. Cooper and Gun's., Tutorial Pharmacy, 6th Edition, Edited by **S. J. Carter.,** page No. 234 – 236.
2. Introduction to chemical engineering, 7th Edition, by **Walter L. Badger and Juliust T. Banchero** page No.577 – 579.
3. Pharmaceutical Engineering principles and practices, 6th Edition, by **C.V.S. Subrahmanyam**, page No. 183 – 189.
4. Pharmaceutical Engineering, 2003 1st Edition, by **K. Sambamurthy,** page No. 248 – 250.
5. Unit operations of chemical engineering, 6th Edition, by **M.C. Cabe, Smith, harriott,** page no. 1003 – 1004.
6. The theory and practice of Industrial Pharmacy, 3rd Edition, by **Leon Lachman, Herbert A. Lieberman,** page no: 146 – 147.

Experiment No : 13 **Dt :**

Determination of Humidity of Air (By Dew Point and Psychrometric Methods)

Aim : To determine the humidity of air by using dew point & psychrometric methods.

Requirements : Psychrometer,

Thermometers,

Cotton, Copper vessel,

Humidity chart, etc

Principle : Humidity is defined as pounds of water vapour carried by one pound of dry air under any given set of conditions. The temperature at which air becomes saturated is the wet bulb temperature. At this temperature air is equilibrium with water.

In dew point method, the dew is found on the walls of container at the temperature, where the air is in contact with it, and gets saturated.

In psychrometric method both the dry bulb and wet bulb temperatures are determined. By determining the wet bulb temperature, humidity of dry air and relative humidity can be calculated with help of humidity chart.

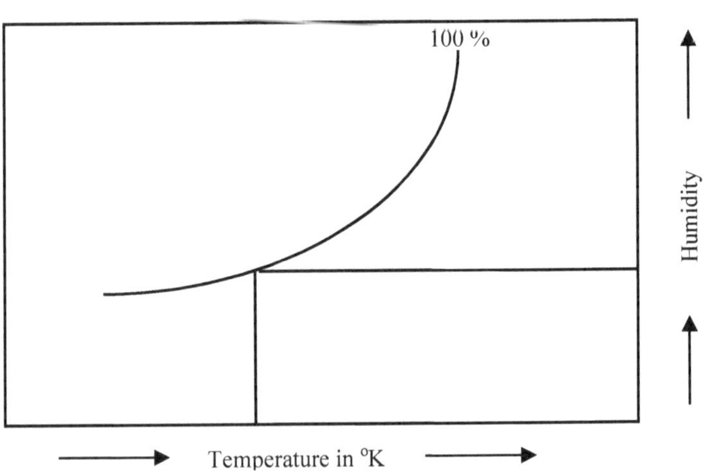

Procedure :

Dew Point Method : Take a copper vessel and fill it with water up to 1/3rd of the vessel. The walls of the vessel should be maintained dry. Add crushed ice pieces slowly and stir continuously until the ice melts. Add the ice pieces and continue the process gradually until mist just starts forming on the walls of the copper vessel. Note the temperature with help of a thermometer. This point is called "***dew point***".

Psychrometric Method : In this method, collect psychrometer & note the temperature in the thermometer initially. It is the dry bulb temperature. Now pack the bulb of one of the thermometer with wet cotton and rotate the psychrometer until there is no further change in the temperature of the thermometer, which is packed with wet cotton. This is noted as the wet bulb temperature. Using humidity chart, the dry bulb humidity is obtained and then relative humidity is calculated by using the following equation :

$$\mathbf{W_w - W_G = (h_G/k_G M_G P)(1/\lambda_w)(t_G - t_w)}$$

Where, W_w = Humidity of Wet bulb temperature

W_G = Humidity of Dry bulb temperature

From exp. measurements, for water $(h_G/k_G M_G P) = 0.26$

λ_w = Latent heat of water at wet bulb temperature

t_w = Wet bulb temperature

t_G = Dry bulb temperature

Observations :

By Dew Point Method :

	Initial temperature		**Dew point temperature**	
S. No	**In ^{o}C**	**In ^{o}F**	**In ^{o}C**	**In ^{o}F**
1				
2				
3				
4				
5				

Report :

Bibliography :

1. The theory and practice of Industrial Pharmacy, 3rd Edition, by **Leon Lachman, Herbert A. Lieberman,** page no : 27.
2. Introduction to chemical engineering, 7th Edition, by **Walter L. Badger and Juliust T. Banchero** page No.577 – 579.
3. Pharmaceutical Engineering, 2003 1st Edition, by **K. Sambamurthy,** page No. 197 – 204.
4. Pharmaceutical Engineering principles and practices, 6th Edition, by **C.V.S. Subrahmanyam**, page No. 183 – 189.
5. Unit operations of chemical engineering, 6th Edition, by **M.C. Cabe, Smith, harriott,** page no. 1003 – 1004.

Experiment No : 14 **Dt :**

Determination of Crystallization

Aim : To observe the effects of various factors on the nature of crystal growth of a super saturated solution by different methods

Requirements : Beakers,

Any salt of potassium (KNO_3 or KCl),

Glycerin, Urea, NaOH and water

Principle : Crystallization is a Unit operation, in which a solid is formed in its pure form from a homogeneous phase of solid either in liquid phase or in a vapour phase. Crystallization mainly depends on *three* (3) conditions,

1. A state of super saturation
2. Formation of nuclei
3. Growth of crystals

Depending on the condition of crystallization, it is possible to control or modify the nature of crystals obtained. The habit or shape of crystal form is highly depending on impurities in solution, pH, rate of cooling and solvent used.

Crystallization is mainly achieved by "super saturation". A super saturated solution can be obtained by different methods.

(i) Super saturation by cooling method

(ii) Super saturation by evaporation

(iii) Super saturation by adiabatic evaporation (Cooling Evaporation)

(iv) Super saturation by salting out effect

If the solubility of a solvent depends on temperature i.e., if the solubility increases with an increase in temperature, super saturation can be brought about by cooling, leading to crystallization.

Adiabatic cooling + evaporation are used for large-scale production.

In salting out method, addition of third substance reduces the solubility of solute up to the point where it crystallizes out.

Procedure :

Preparation of super saturated solution : Super saturated solution is one that contains more of the dissolved solute than it would normally contain at a definite temperature where the un-dissolved solute present.

Prepare a saturated solution of a given salt by adding excess amount of salt in water (more than the solubility point), and decant the solution. Then collect the supernatant liquid and proceed for further steps.

Crystallization by Cooling : Saturated solution of salt is taken at room temperature and divides into two portions in equal proportions and keeps in an ice bath. One of the containers is kept constantly without any shaking (disturbance) and the other is shaken continuously, till the growth of crystals observed in both containers. Note down the time and crystal shape, nature of the crystals observed microscopically.

Salting-Out Method : Saturated solutions of salt is taken at room temperature and divides into three portions in equal proportions and add glycerin in one of the three portions, urea in one of the remaining two portions, NaOH in last portion of the remaining, and keep the containers aside for some time to form crystals. Then observe the crystal growth, and note down the time, shape of the crystals.

Take saturated solution and add glycerin with agitation and observe the growth of crystals, and determine the nature of the crystal growth under microscopically.

Observation :

Method of Crystallization	Time taken	Nature / Shape of crystals
Cooling without agitation		
Cooling with agitation		
Crystal form with Urea		
Crystal form with NaOH		
Crystal form with Glycerine		
Crystal form with Glycerine with agitation / stirring		

Report :

Bibliography :

1. Cooper and Gun's, Tutorial Pharmacy, 6th Edition, Edited by **S. J. Carter.,** page No. 234 – 236.
2. Introduction to chemical engineering, 7th Edition, by **Walter L. Badger and Juliust T. Banchero** page No.577 – 579.
3. Pharmaceutical Engineering, 2003 1st Edition, by **K. Sambamurthy,** page No. 248 – 250.
4. Pharmaceutical Engineering principles and practices, 6th Edition, by **C.V.S. Subrahmanyam**, page No. 183 – 189.

Experiment No : 15 **Dt :**

Simple Distillation

Aim : To separate the pure liquid from a given impure liquid mixture.

Requirements : Distillation flask (still),
Condenser, L – bend tube & Receiver

Principle : Simple distillation is a process of converting a single constituent from a liquid mixture into its vapour, transferring the vapour to another place and recovering the liquid by condensing the vapour, usually by allowing it to come in contact with a cold surface.

Liquid boils when its vapour pressure is equal to atmospheric pressure. Simple distillation is conducted at its boiling points. The higher the relative volatility of a liquid, the better is the separation by a simple distillation. Heat is supplied to the liquid so that it boils. The resulting vapour is transferred to a different place and condensed. If the liquid of interest is volatile and remaining components are non-volatile, then simple distillation is a useful means of purification and separation of liquids.

The efficiency of distillation is calculated by using the following equation;

$$\% \text{ Efficiency of distillation} = \frac{\text{Practical yield}}{\text{Theoretical yield}} \times 100$$

Theory : The purification of many organic liquids is carried out by the distillation process and it is necessary to use only a simple apparatus with a boiler and a condenser, small scale laboratory units are made of glass and a typical setup is shown in figure. The essential parts of a simple distillation unit are a still where the liquid is boiled, a condenser where the vapours are condensed to liquid and a receiver to connect the condensed liquid. To minimize condensation in the flask, its upper part should be well insulated from heat loss with a material like asbestos rope or well covered with a dry cloth.

Procedure :

1. Collect the required apparatus and assemble them as shown in figure.
2. Take the liquid mixture, which is to be distilled and filled into the flask up to $1/3^{rd}$ of its volume.
3. Heat the apparatus to the temperature not more than (NMT) 110 °C (If the R.B. Flask is kept without water bath).
4. Adding small pieces of porcelain or porous pot or bumping pumice stone before distillation, avoids bumping.
5. The liquid start vaporizes, distilled and the vapors are circulated through the jacket of the condenser and condensed.
6. Distillate is collected in a collector / receiver.
7. Finally, measure the volume of the distilled liquid and calculate the percentage efficiency of distillation.

Applications :

1. Simple distillation is used for the preparation of distilled water & water for Injection.
2. It is used for the preparation of volatile & aromatic waters.
3. It is used for the purification of organic solvents.
4. It is used for the separation of volatile liquids from the non-volatile liquids.
5. A few official compounds (ex: Spirit of nitrous ether, Aromatic spirit of ammonia) are prepared by distillation.

Observations :

$$\%\ \text{Efficiency of distillation} = \frac{\text{Practical yield}}{\text{Theoretical yield}} \times 100$$

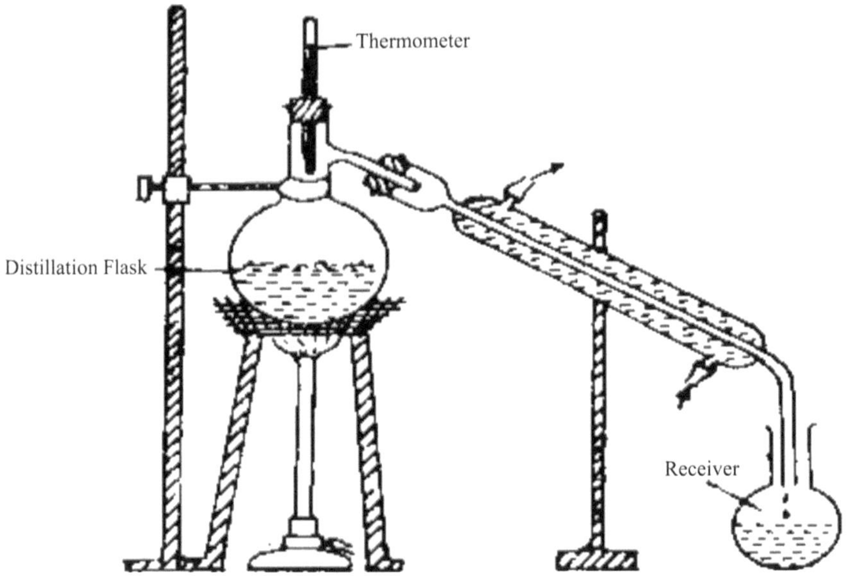

Fig. 15.1 Experimental setup of simple **Distillation.**

Report :

Bibliography :

1. Cooper and Gun's., Tutorial Pharmacy, 6th Edition, Edited by **S. J. Carter.,** page No. 262 – 263.
2. Introduction to Chemical engineering, 7th Edition, by **Walter L. Badger and Juliust T. Banchero** page No.245 – 258.
3. Pharmaceutical Engineering, 2003 1st Edition, by **K. Sambamurthy,** page No. 134 – 137.
4. Pharmaceutical Engineering principles and practices, 6th Edition, by **C.V.S. Subrahmanyam**, page No. 304 – 308.

Experiment No : 16 **Dt :**

Steam Distillation

Aim : To find out the efficiency of steam distillation process for distillation of high boiling solvents from a non–volatile substance (liquid mixture).

Requirements : Distillation flask, Thermometer, Condenser, Bend tubes, Steam generator, Double holed rubber cork

Principle : *Steam distillation* is a method of distillation carried with the aid of steam and is used for the separation of high–boiling substances from non–volatile impurities. A mixture of immiscible liquids begins to boil when the sum of their vapour pressures is equal to the atmospheric pressure. In case of a mixture of turpentine & water, the turpentine boils below the temperature than that of pure water boiling point.

Steam distillation is especially used for the separation of volatile oils at low temperature without decomposition.

For example, the boiling point of turpentine is about 160 °C. But when it is mixed with water and heated, the mixture boils at about 95.6 °C. At this temperature, the vapour pressure of water is 86.245 kPa (647 mm Hg) and that of turpentine is 15.06 kPa (113 mm Hg). The sum of the vapour pressures is 101.31 kPa (760 mm Hg), which is normal atmospheric pressure. Thus, high boiling substances may be distilled at a temperature much below its boiling point, when water (steam) is used.

The efficiency of distillation is calculated by using following equation;

$$\%\ \text{Efficiency of distillation} = \frac{\text{Practical yield}}{\text{Theoretical yield}} \times 100$$

Procedure :

1. On laboratory scale, the distillation apparatus consists of steam generator fitted with a rubber cork having two holes.
2. Through one hole a long safety tube is passed which permits the expulsion of some water, if excessive pressure is generated inside the steam generator.
3. Escape of steam from the safety tube indicates that, the steam can is almost dry.
4. Through the other hole, a bent tube is passed which carries the steam to the bottom of the flask.
5. The steam generator and flask containing the liquids are heated simultaneously, so that a rapid current of steam passes through the boiling mixture in the flask.
6. The liquid begins to boil when the sum of the two vapour pressures equal to atmospheric pressure.
7. The vapours are allowed to pass through the condenser and condensed liquids are collected in Florentine receiver. The distillate, which forms two layers, one aqueous and other as completely as possible.
8. For this purpose either Florentine receivers or separating funnels are used.

Applications :

1. Steam distillation is used for the separation of immiscible liquids. Ex: toluene & water.
2. It is used for extracting most of the volatile oils such as clove, anise and eucalyptus.
3. It is useful in purification of liquid with high boiling point. *Ex :* Essential oil of almond.
4. Camphor is distilled by this method.
5. Aromatic waters are prepared by this method.

Observations :

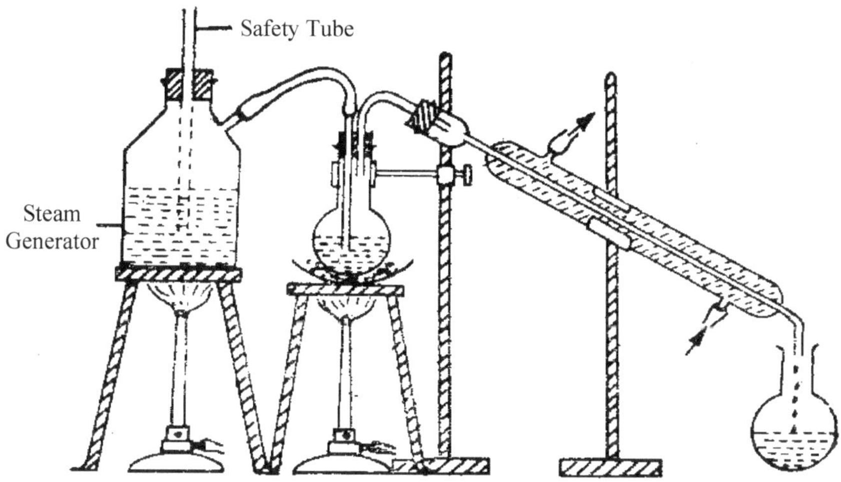

Fig. 16.1 Experimental setup of steam distillation.

Report :

Bibliography :

1. Introduction to chemical engineering, 7th Edition, by **Walter L. Badger and Juliust T. Banchero** page No.577 – 579.
2. Pharmaceutical Engineering, 2003 1st Edition, by **K. Sambamurthy,** page No. 152 – 154.
3. Pharmaceutical Engineering principles and practices, 6th Edition, by **C.V.S. Subrahmanyam**, page No. 304 – 306.

Experiment No : 17 **Dt :**

Verification of Stoke's Law

Aim : To verify the stoke's law for a given sphere (solid material) in a given liquid.

Requirements : Liquid medium (glycerin), Spheres (metal balls of different sizes), stop clock, screw gauze

Principle : In free settling, the particles settle down independently of one another. The particles will accelerate under the influence of gravity. As the particles accelerate, fluid offers a greater and greater fractional resistance. A stage will be reached when the resistance force is exactly equal to force of gravity, then the acceleration of particles become zero, and it will settle at a definite constant velocity from the time. The velocity is called "*Terminal settle velocity*".

The various factors involved in the rate velocity of settling, the particles of a suspension are embodied in the equation of ***stoke's law***.

$$V = \frac{2r^2(\rho_1 - \rho_2)g}{9\eta} \quad \text{Or} \quad V = \frac{d^2(\rho_1 - \rho_2)g}{18\eta}$$

Where, V = the velocity of the fall of spherical particle (ball)

g = gravitational constant (980.7 cm/sec^2)

d = diameter of the sphere (in cm)

r = particle radius (in cm)

ρ_1 = density of the sphere (gm/ml)

ρ_2 = density of the liquid (gm/ml)

η = viscosity of the dispersion medium (Poises)

Density of sphere can be calculated by **ρ = Mass/Volume**

Volume of sphere can be calculated by using following formula;

$$V = \frac{\pi d^3}{6}$$

Water at 20 °C has a viscosity of approximately 1 cp. (0.01 Poise)

Stoke's law was derived for dilute suspension of rigid uniform spheres settling at a velocity which produced no turbulence in the fluid medium. Under much conditioned the particles do not interfere with each other that is fast settling occurs.

Increase in the mean particle size or increase in the difference between the densities of the solid and liquid phases will produce a faster rate of sedimentation where an increase in the viscosity of liquid medium will decrease the rate.

Limitations :

1. Pharmaceutical systems containing less than 2 gm of solids per 100 ml generally follow stoke's equation.
2. Stoke's equation was derived for an ideal situation in which uniform, perfectly spherical particles in a very dilute suspension settle without effecting turbulence in their downward course, without collision of the particles of the suspension, and without chemical or physical attraction or affinity for the dispersion medium.
3. Stoke's equation does not apply precisely for the suspension is irregularly shaped, of various particle diameters & not spherical.
4. If the particles are subjected only to the force of gravity. Then the lower size limit of particles obeying stoke's equation is about 0.5 μm.
5. The particles must not be aggregated for clamped together into fluid.
6. Stoke's law undergoes only for the sphere falling freely without hinderable and at a constant rate.

Procedure :

1. Take a long transparent glass tube, which is marked A and B in with different distance & filled with glycerin carefully without forming air bubbles.

2. The tube is fixed to stand in a vertical position without any inclination.
3. Take three different set of spheres and weighed individually and diameter of each is determined by means of screw gauze. The radius is then calculated.
4. Find out the volume of sphere by using the formula, $V = 4/3\ \pi r^3$ (or) $V = \pi d^3/6$, and then **density** is calculated by dividing **mass** by **volume (ρ = m/v)**.
5. One sphere from 1st set is taken and drops it on the top of the glycerin glass tube. Make sure that, it should pass through the middle of the tube.
6. Stop clock is switched on when the sphere is passed through **mark A** of the glycerin and switch off when it reached **mark B**.
7. Note down the time taken for travel the sphere from **mark A** to **mark B** and distance (height) between **mark A** and **mark B**.
8. Repeat this procedure for remaining spheres also.
9. Then, the velocity is calculated by using the formula, V = d/t (distance/time).
10. The practical velocity compared with the theoretical velocity obtained by stoke's law.

Applications :

1. Used for deflocculated systems, where in particles settle independently.
2. It is useful in quantitative analysis, which can be utilized in formulation of suspensions.
3. Stoke's law is applicable for dilute pharmaceutical suspensions (2 gms of solids per 100 ml)

Observation :

Stoke's Law

$$V = \frac{d^2(\rho_1 - \rho_2)g}{18\eta}$$

S. No	Wt. (gms)	Dia-meter (cms)	Radius (cms)	Volume ($V=4/3\pi r^3$)	Density of Sphere (ρ=w/v)	Time taken to pass through mark A to B	Practical velocity (V = d/t)	Theoretical Velocity (Stoke's law)

Bibliography :

1. Introduction to chemical engineering, 7th Edition, by **Walter L. Badger and Juliust T. Banchero** page no.637 – 638.
2. Physical Pharmacy, 6th Edition, by **C.V.S. Subrahmanyam,** page no. 374 – 375.
3. Physical Pharmacy, 3rd Edition; by **Alfred Martin** page no. 547 & 505 – 506.
4. Introduction to Pharmaceutical dosage forms, 1st Edition, by **C.Ansel** page no. 209 – 210.
5. The theory and practice of Industrial Pharmacy, 3rd Edition, by **Leon Lachman, Herbert A. Lieberman,** page no: 484 – 485.

Experiment No : 18 **Dt :**

Sedimentation by Centrifugation

Aim : To estimate the sedimentation time for the formulation, having different concentrations of suspending agents by using centrifugation.

Requirements : Calamine lotion prepared with different concentrations of suspending agents, centrifuge, scale, stop clock, etc.

Principle : Centrifugal force is used to provide the derived force, to replace the gravitational force in sedimentation process. In this experiment, the relative sedimentation of four aggregate material (Bentonite) concentrations [3%, 2%, 1% and 0%(Control) respectively–calamine lotion] using a centrifuge is compared. Sedimentation by centrifugation involves utilization of centrifugal force for sedimentation of solids. During the process of sedimentation, coarse material falls to bottom above which, layer of settle solid with transition zone of partly thickened material above it. The boundary between transition zone & settled solid region is usually observe & through this water escapes from lower layer which are under compression. Above transition zone, pulp at original concentration & layer of clear water are found. As thickening progress due to sedimentation the top two layers disappear & layer of settled solids shrink, because of compression. Thus as the time of sedimentation increases, the height of sediments of prepared concentration of suspension decrease to some extent after with the weight remains constant. At this time count sediments are formed.

Theory : Centrifuging (also called Centrifugation) is a common unit operated in pharmacy. Centrifugal force commonly expressed in multiples of the force of gravity, varies with the rotational speed and with the radial distance from the center of rotation.

A centrifuge is an apparatus utilizing centrifugal force for the separation of liquid from solids. It is essentially a development of gravity filter wherein the force acting on the liquid, instead of being restricted to gravity, is enormously increased by utilizing centrifugal force. This increased force can also be applied to the separation of immiscible liquids.

Sedimentation Centrifuge : A centrifuge that produces sedimentation of solids based on the difference in the densities of two or more phases of the mixture.

Ex : Top–suspended centrifuges are used extensively in sugar refining, automatic batch centrifuges, and continuous filtering centrifuges, etc.

Procedure :

1. Prepare four 10ml calamine lotions by using 0.0%, 1%, 2% and 3% suspending agent (Bentonite) respectively.
2. Transfer into the centrifuge tube & measure the height of sediment in each tube.
3. Operate centrifuge and determine height of sediment after every 5 min.
4. Take 6 (six) readings or until constant heights obtained.
5. Plot a graph between, **height of sediment** on Y-axis & **time** on X-axis.

Observation :

Preparation of Calamine Lotion :

INGREDIENTS	For 100 ml	For 10 ml
Calamine	15 gm	1.5 gm
ZnO	5 gm	0.5 gm
Bentonite	-	-
Sodium Citrate	0.5 gm	0.05 gm
Liquefied Phenol	0.5 ml	0.05 ml
Glycerin	1 ml	0.1 ml
Water	Up to 100 ml	Up to 10 ml

Table for Observing Sediment of Height :

	Sediment height		
Time in min.	**3%**	**2%**	**1%**
0			
5			
10			
15			
20			
25			
30			
35			

Graph : A graph between **Sediment heights** on Y-axis Vs **Time (in min)**. on X-axis

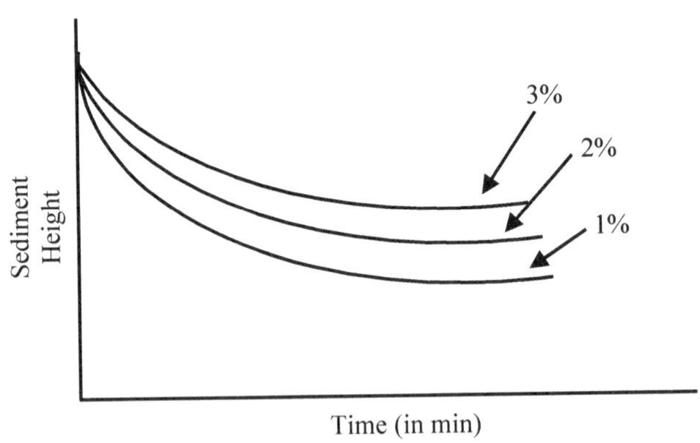

Report :

Bibliography :

1. Introduction to chemical engineering, 7th Edition, by **Walter L. Badger and Juliust T. Banchero** page no.593 – 599.
2. Pharmaceutical Engineering, 2003 1st Edition, by **K. Sambamurthy,** page no. 282 – 286.
3. Unit operations of chemical engineering, 6th Edition, by **M.C. Cabe, Smith, harriott,** page no. 1000 – 1003.
4. Physical Pharmacy, 3rd Edition; by **Alfred Martin.,** page no. 547 – 548.
5. The theory and practice of Industrial Pharmacy, 3rd Edition, by **Leon Lachman, Herbert A. Lieberman,** page no. 484 – 485.

Experiment No : 19 **Dt :**

Overall Heat Transfer Coefficient (Counter Current Flow)

Aim : To determine overall heat transfer coefficient for a given heat interchanger from a given flask of glass using counter current flow.

Requirements : Distillation flask, Thermometer, Condenser, Bend tubes, Steam generator, Double holed rubber cork

Principle : Consider a heat interchanger, a hot fluid inside a pipe is cooled from T_1 to T_2 by transferring its heat to a cold fluid outside the pipe, entering at t_1 and heated to t_2. Liquid is heated with steam using distillation setup, a temperature gradient exists through the wall between the steam of liquid when steam is condensed it moves through bulb of liquid. The heat is transferred through conduction via film and the surface. The rate of heat transfer is the product of overall heat transfer coefficient area of heating surface & temperature drop.

The overall coefficient may be considered constant throughout the experiment, and specific heat of each fluid may be considered constant, it can be shown

$$\mathbf{Q = UA\ \Delta T_m}$$

Where, Q = rate (quantity) of heat transfer

A = area of heating surface

ΔT_m = mean temperature difference,

U = over-all heat transfer coefficient

$\Delta t_m = (\Delta t_2 - \Delta t_1)/2.303 \log (\Delta t_2/\Delta t_1)$

$$\mathbf{U = Q/A\ \Delta T_m}$$

$$Q = q_1 + q_2$$

q = heat loss by steam, $q_1 = m_1.s_1.\Delta T_1 + L.m_1$

$$\mathbf{q_2 = m_2.s_2.\Delta T_2}$$

Where, m_1 = mass of condensate

S_1 = specific heat of condensate

L = latent heat of vaporization of steam (540 cal./A^o)

m_2 = mass of circulating water,

S_2 = specific heat of circulating water

$\Delta T = (t_2 - t_1)$ temperature drop

Where, t_2 = temperature of condensate,

t_1 = temperature of outlet

Theory : The temperature gradients in this case are shown in following figure. It can be seen that the temperature drop along the length of the apparatus is much more nearly constant than in the case of parallel flow and hence transfer is more nearly constant throughout the apparatus. Also in counter current operation the exit temperature of the hot fluid can be considerably less than the exit temperature of the cold fluid, and accordingly a larger proportion of heat content of the hot fluid can be extracted for a given entrance temperature of the cold fluid.

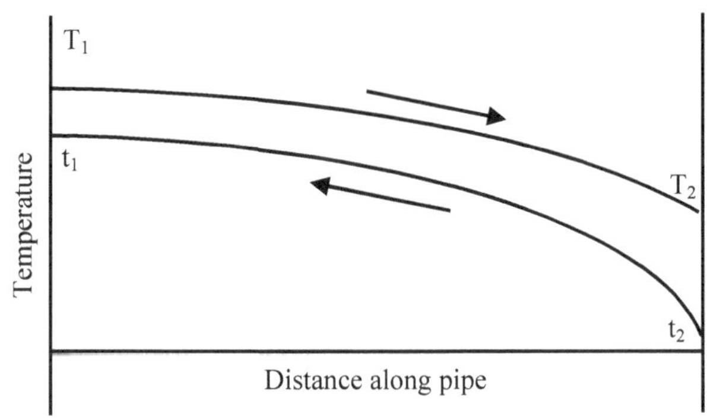

Procedure :

1. Take all required apparatus and arrange as setup,
2. Flame it and generated steam is passed through water–jacketed glass container.
3. It acts as a heat interchanger. Circulating cold water in parallel flow manner condenses the steam.
4. Each time temperature of inlet and outlet water, volume of circulating water are determined for a definite period of time.
5. Mass of condensate to be determined, length and diameter of condenser are determined.

6. Experiment is repeated by varying speeds of circulating water.
7. In each case over-all heat transfer coefficient is calculated.

Note : Start collecting the water when the condensate begins to form

Precautions :

1. Covering the condenser must prevent heat loss through radiation.
2. Care should be taken to avoid escape of steam through any outlet.

Observation :

With Low Speed :

S. No	Temp. of inlet water	Temp. of inlet water	Temp. of condensate	Vol. of condensate	Vol. of outlet water
1					
2					
3					
Average					

With Medium Speed :

S. No	Temp. of inlet water	Temp. of inlet water	Temp. of condensate	Vol. of condensate	Vol. of outlet water
1					
2					
3					
Average					

With Low & Medium Speeds :

$$q_1 = m_1.s_1\Delta T_1 + m_1.L\, q_2 = m_2.s_2\Delta T_2$$

Average steam temperature = 100 °C

Average inlet water temperature (Room temperature) =

Temperature drop between outlet and inlet water (ΔT_1) =

Temperature drop between steam and inlet water (ΔT_2) = 100 °C – R.T (room temp).

Area of heating surface = $2\pi r.l$

Circumference of heating surface (C) = $2\pi r$

Length of condenser (l) = cm

$$\mathbf{Q = q_1 + q_2}$$

$$\mathbf{U = Q/A.\Delta T_m}$$

Report :

Bibliography :

1. Introduction to chemical engineering, 7th Edition, by **Walter L. Badger and Juliust T. Banchero** page no.143 – 144.
2. Pharmaceutical Engineering, 1st Edition 2003, by **K. Sambamurthy,** page no. 74–75.
3. Unit operations of chemical engineering, 6th Edition, by **M.C. Cabe, Smith, harriott,** page no. 315 – 319.
4. Pharmaceutical Engineering principles and practices, 6th Edition, by **C.V.S. Subrahmanyam**, page No. 125 – 126.

Experiment No : 20 **Dt :**

Overall Heat Transfer Coefficient (Parallel Flow)

Aim : To determine overall heat transfer coefficient for a given heat interchanger from a given flask of glass using counter current flow.

Requirements : Distillation flask, Thermometer, Condenser, Bend tubes, Steam generator, Double holed rubber cork

Principle : Consider a heat interchanger, a hot fluid inside a pipe is cooled from T_1 to T_2 by transferring its heat to a cold fluid outside the pipe, entering at t_1 and heated to t_2. Liquid is heated with steam using distillation setup, a temperature gradient exists through the wall between the steam of liquid when steam is condensed it moves through bulb of liquid. The heat is transferred through conduction via film and the surface. The rate of heat transfer is the product of overall heat transfer coefficient area of heating surface & temperature drop.

The overall coefficient may be considered constant throughout the experiment, and specific heat of each fluid may be considered constant, it can be shown

$$\mathbf{Q = UA\,\Delta T_m}$$

Where, Q = rate (quantity) of heat transfer

A = area of heating surface

ΔT_m = mean temperature difference,

U = over-all heat transfer coefficient

$$\Delta T_m = (\Delta t_2 - \Delta t_1)/2.303 \log (\Delta t_2/\Delta t_1)$$

$$\mathbf{U = Q/A\,\Delta T_m}$$

$$\mathbf{Q = q_1+q_2,}$$

q = heat loss by steam, $q_1 = m_1.s_1.\Delta T_1 + L.m_1$

$$\mathbf{q_2 = m_2.s_2.\Delta T_2}$$

Where, m_1 = mass of condensate

S_1 = specific heat of condensate

L = latent heat of vaporization of steam (540 cal./A°)

m_2 = mass of circulating water,

S_2 = specific heat of circulating water

$\Delta T = (t_2 - t_1)$ temperature drop

Where, t_2 = temperature of condensate,

t_1 = temperature of outlet

Theory : The temperature gradients in this case are shown in following figure. It can be seen that the temperature drop along the length of the apparatus is much more nearly constant than in the case of parallel flow and hence transfer is more nearly constant throughout the apparatus. Also in counter current operation the exit temperature of the hot fluid can be considerably less than the exit temperature of the cold fluid, and accordingly a larger proportion of heat content of the hot fluid can be extracted for a given entrance temperature of the cold fluid.

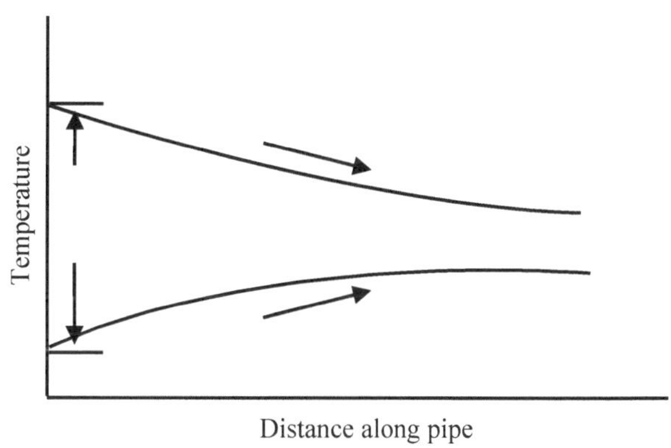

Procedure :

1. Take all required apparatus and arrange as setup,
2. Flame it and generated steam is passed through water–jacketed glass container.
3. It acts as a heat interchanger. Circulating cold water in counter current flow manner condenses the steam.
4. Each time temperature of inlet and outlet water, volume of circulating water are determined for a definite period of time.

5. Mass of condensate to be determined, length and diameter of condenser are determined.
6. Experiment is repeated by varying speeds of circulating water.
7. In each case over-all heat transferred coefficient is calculated.

Note : Start collecting the water when the condensate begins to form

Precautions :

1. Covering the condenser must prevent heat loss through radiation.
2. Care should be taken to avoid escape of steam through any outlet.

Observation :

With Low Speed :

S. No	Temp. of inlet water	Temp. of inlet water	Temp. of condensate	Vol. of condensate	Vol. of outlet water
1					
2					
3					
Average					

With Medium Speed:

S. No	Temp. of inlet water	Temp. of inlet water	Temp. of condensate	Vol. of condensate	Vol. of outlet water
1					
2					
3					
Average					

With Low & Medium Speeds :

$$q_1 = m_1.s_1\Delta T_1 + m_1.L q_2 = m_2.s_2\Delta T_2$$

Average steam temperature = 100 °C

Average inlet water temperature (Room temperature) =

Temperature drop between outlet and inlet water (ΔT_1) =

Temperature drop between steam and inlet water (ΔT_2) = 100 °C – R.T (room temp).

Area of heating surface = 2πr.l cm^2

Circumference of heating surface (C) = 2πr cm

Length of condenser (l) = cm

$$\mathbf{Q = q_1 + q_2}$$

$$\mathbf{U = Q/A.\Delta T_m}$$

Report :

Bibliography :

1. Introduction to chemical engineering, 7th Edition, by **Walter L. Badger and Juliust T. Banchero** page no.142 – 143.
2. Pharmaceutical Engineering, 1st Edition 2003, by **K. Sambamurthy,** page no. 74 – 75.
3. Unit operations of chemical engineering, 6th Edition, by **M.C. Cabe, Smith, harriott,** page no. 315 – 319.
4. Pharmaceutical Engineering principles and practices, 6th Edition, by **C.V.S. Subrahmanyam**, page No. 124 – 125.

Experiment No : 21 **Dt :**

Identification of Azeotropic Mixtures

Aim : To identify the azeotropic mixture with the different techniques of distillation method.

Requirements : Distillation flask, Thermometer, Condenser, Bent tubes, Steam generator, Double holed rubber cork

Theory : The word *AZEOTROPE* comes from the Greek "*zein tropos*" or "Constant boiling". An azeotrope said to be **positive** when the constant boiling point is at a temperature maximum and **negative** when the boiling point is at a temperature minimum. The vast majority of azeotropes are minimum boiling. All liquid mixtures which are immiscible and which form azeotropes are minimum boiling point. Separation of the original mixture may also be enhanced by adding a solvent that forms as azeotrope with one of the key components. This process is called azeotropic distillation. Usually the material added forms a low boiling azeotrope and is taken overhead, and such materials are called "*Entrainers*".

Ex : Use of benzene, heptane, or cyclohexane to permit the separation of ethanol and water, which forms a minimum boiling azeotrope with 95.6 % weight alcohol. The alcohol-water mixture with about 95 % alcohol is fed near the middle of the dehydrating column, and nearly pure alcohol is removed as the bottoms product (Alcohol *BP* 78.3°C and water *BP* 100°C form a minimum boiling azeotrope, *BP* 78.15°C containing 95.57% w/w alcohol. Addition of benzene to this binary azeotrope as the entrainer and distillation results in the formation of a ternary azeotrope containing 18.5% alcohol, 7.4% water and 74.1% benzene with *BP* 64.85°, which is lower than the *BP* of any of the components). The overhead vapour is a ternary azeotrope, which is condensed and separated into two phases. The upper organic layer is returned to the top of the dehydrating column as reflux, and the water taken overhead and returned to the first column. The stripper bottom is an aqueous stream that is discharged as wastewater or sent to a third column to recover some of the alcohol. Azeotropic mixtures are of two types.

Minimum boiling point azeotropic solution : This is a type of azeotropic mixture with a maximum vapour pressure or minimum boiling point. These systems are known as type II solutions. Ex; Chloroform & water; water & nitric acid

Maximum boiling point azeotropic solution : This is a type of azeotropic mixture with a maximum vapour pressure or maximum boiling point. These systems are known as type III solutions. Ex; water & ethanol; benzene & ethanol

There are several procedures for breakup of the azeotrope.

1. Use azeotropic distillation, by adding in a third component, which will generate a lower boiling heterogenous azeotrope that, can be broken by phase separating the two immiscible liquids and decanting.
2. Use extractive distillation where a third compound is added near the top of a distillation column that reduces the volatility of one compound over another.
3. Use liquid-liquid extraction to separate the compounds by segregating them into two different liquid phases.
4. Use a membrane to separate them, as in pervaporation.
5. For azeotropes that change composition when pressure is changed, you can use pressure swing distillation by operating two distillation columns at different pressures, and recycle material between these two distillation columns. One column will remove a pure component (two-component systems) in the bottom of one distillation column, the other will remove a pure component in the bottom of a second distillation column, and the azeotrope (of two different compositions) will be fed to the other column.

Principle : An azeotrope is defined as a liquid mixture of two or more liquid substances that remains the same composition in the vapour state as in the liquid state when distilled or partially evaporated under a certain pressure (or) An azeotrope is defined as a mixture of two or more liquid components, the boiling point of which does not change as vapour is generated and removed. These are also called *constant-boiling mixtures*. The azeotropic mixture has a lower boiling point. At the minimum boiling point temperature, the liquid composition remains constant and is equal to the vapour composition. Azeotropes which exist as one liquid phase in equilibrium with vapour may be called homogeneous azeotropes, as distinguished from those existing as two liquid phases in equilibrium with vapour which may be called heterogeneous azeotropes.

Ex : 2-Propanol / Water, Nitric acid (68%) / Water, Ethanol/Water, etc

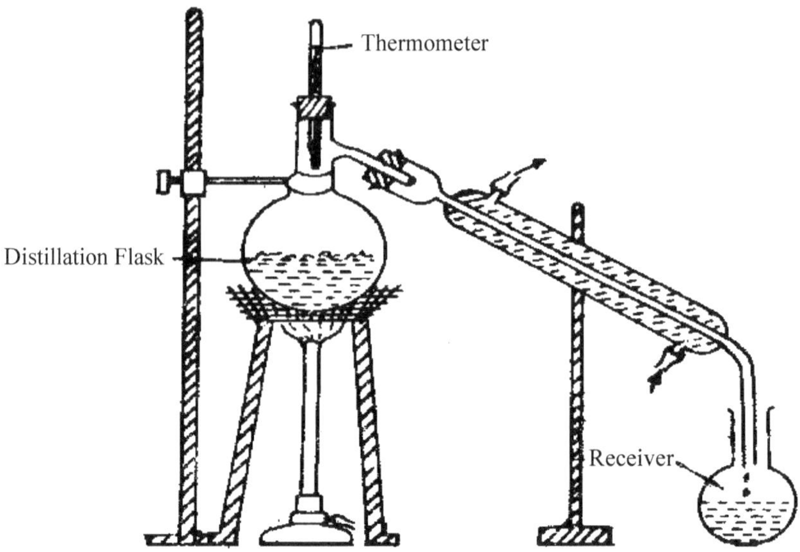

Fig. 21.1 Experimental Setup For Constant Boiling Point Mixture.

Procedure :

1. Collect the clean and dried apparatus and assemble them as shown in figure.
2. Take the azeotropic mixture (Ex: ethanol & water) and filled into the flask up to $1/3^{rd}$ of its volume.
3. Adding small pieces of porcelain or porous pot or bumping pumice stone before distillation, avoids bumping.
4. Heat the apparatus to the temperature up to its boiling point.
5. Once the liquid start to vapourize, the vapours passes through the condenser and condenses.
6. Collect the distillate with the help of collector / receiver.
7. If this mixture boils continuously for long time at the same temperature, (which is shown in thermometer), then this point is called as Constant boiling temperature. This indicates that the mixture is *Azeotropic mixture*.

Applications :

1. The liquor from fermentation process is a common source of ethanol in a concentration of approximately 8 to 10%.
2. Absolute alcohol can be prepared by azeotropic distillation.

Examples of Azeotropes

- nitric acid (68%) / water, boils at 120.5°C at 1 atm
- perchloric acid (28.4%) / water, boils at 203°C (negative azeotrope)
- hydrofluoric acid (35.6%) / water, boils at 111.35°C (negative azeotrope)
- ethanol (96%) / water, boils at 78.2°C
- sulfuric acid (98.3%) / water, boils at 100°C
- acetone / methanol / chloroform form an intermediate boiling azeotrope
- diethyl ether (33%) / halothane (66%) a mixture once commonly used in anaesthesia.
- benzene / hexafluorobenzene forms a double binary azeotrope.

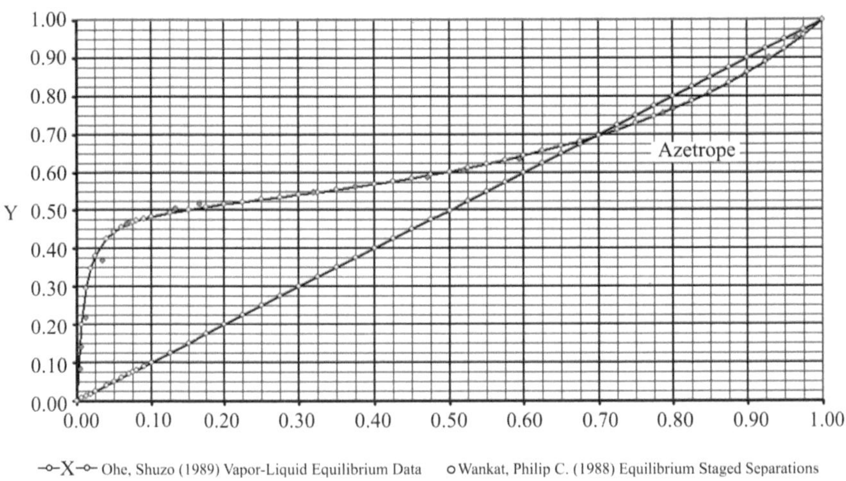

Fig. 21.2 Datos de equilibrio 2 Propanol / agua P = 760 mm Hg Bibliograficos.

Bibliography :

1. Introduction to chemical engineering, 7th Edition, by **Walter L. Badger and Juliust T. Banchero** page no.142 – 143.
2. Pharmaceutical Engineering, 1st Edition 2003, by **K. Sambamurthy,** page no. 74 – 75.
3. Unit operations of chemical engineering, 6th Edition, by **M.C. Cabe, Smith, harriott,** page no. 315 – 319.
4. Pharmaceutical Engineering principles and practices, 6th Edition, by **C.V.S. Subrahmanyam**, page No. 124 – 125.

Experiment No : 22 **Dt :**

Construction of Boiling Point Diagram

Aim : To construct the boiling point diagram for a given set of mixture.

Requirements : Glacial Acetic acid, Distilled flask, condenser, conical flask, burette, beaker, Oxalic acid, Phenolphthalein indicator, etc

Principle : In boiling point diagram, temperatures are plotted as ordinates and composition as abscissa. The diagram consists of two curves and both end of which coincide; any point on the upper curve as its abscissa a composition of vapour. That will first begin to condensate at the temperature given by ordinate T_1 and will give liquid composition X. any point like "d" on lower curve has for its abscissa, the composition of the liquid that will just begin to boil at the composition of liquid & vapour in equilibrium with each other at a temperature given by horizontal line through them. For all points above the top line such as A, the mixture is entirely vapour state. For all points blow B the mixture is liquid state.

The boiling point diagrams explain the way to achieve complete separation of two miscible liquids from their mixtures.

Procedure :

1. Arrange the distillation setup as shown in Fig. 21.1 (Previous Experiment).
2. Prepare different compositions of mixture (Acetic acid & Water) as shown in table:
3. Transfer the acetic acid–water mixture in ratio 40:10 into the distillation flask.
4. Heat the flask and distill until a constant temperature is reached i.e. no further increase in Boiling point of mixture (temperature).
5. At this temperature collect 5 ml of condensate in a small beaker, and simultaneously measure 1 gm of distillate using a weighing bottle.
6. Transfer the 1 gm of distillate into a conical flask and titrate against 1N NaOH using Phenolphthalein indicator.

7. Determine the weight of acetic acid present in the 1 gm sample; from this determine the weight of water, there by mole fraction & % mole fraction.
8. Repeat the entire experiment for the remaining different proportions (30:20, 20:30, 10:40) mixtures.
9. Construct the graph by plotting % mole fraction on X-axis & and Temperature on Y-axis.

Diagram :

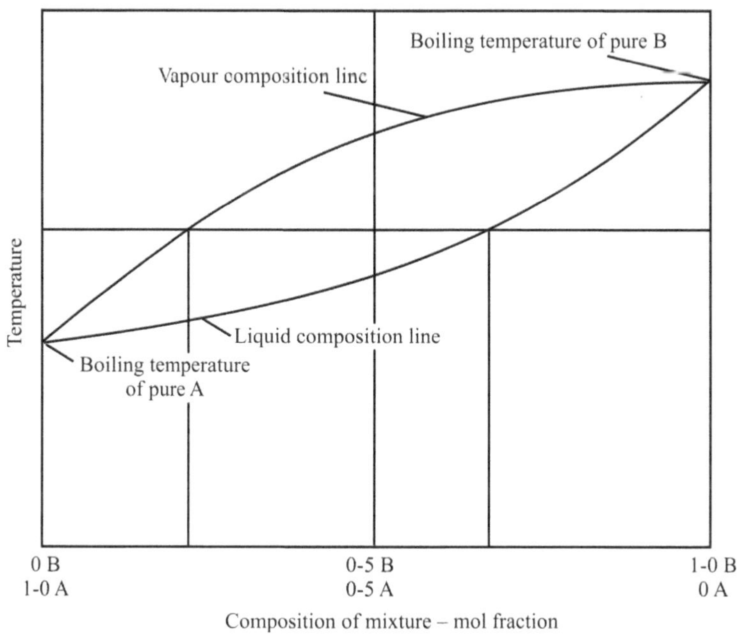

Observation :

1. Number of moles of acetic acid = $\dfrac{\text{Amount of acetic acid}}{60}$

2. Number of moles of Water = $\dfrac{\text{Amount of water}}{18}$

3. Mole fraction of acetic acid = $\dfrac{\text{Moles of acetic acid}}{\text{Moles of acetic acid + Moles of water}}$

4. Mole fraction of water = $\dfrac{\text{Mole of water}}{\text{Moles of acetic acid + Moles of water}}$

Table 22.1 For Liquid and Vapour Phases.

S. No.	Composition of mixture	Boiling Point	Wt. of Sample	Wt. of NaOH	Wt. of A.acid	Wt. of Water	No. of Moles of A.acid	No. of Moles of water	Mole fraction	% Mole fraction
LIQUID PHASE										
1	40:10									
2	30:20									
3	20:30									
4	10:40									
VAPOUR PHASE										
1	40:10									
2	30:20									
3	20:30									
4	10:40									

Report :

Bibliography :

1. Introduction to chemical engineering, 7th Edition, by **Walter L. Badger and Juliust T. Banchero** page no.245 – 247.
2. Pharmaceutical Engineering, 1st Edition 2003, by **K. Sambamurthy,** page no. 127 – 128.
3. Unit operations, 4th Edition, by **George Granger Brown,** page no. 485 – 486.
4. Pharmaceutical Engineering principles and practices, 6th Edition, by **C.V.S. Subrahmanyam**, page No. 311 – 312.

Experiment No : 23 **Dt :**

Efficiency of Single and Multiple Extractions

Aim : To compare the extraction efficiencies of single and multiple stage extractions.

Requirements : Distillation flask, Thermometer, Condenser, Bent tubes, Steam generator, Double holed rubber cork

Principle : The efficiency of extraction in a maceration process can be improved by arranging for the solvent to be continuously circulated through the drug. Solvent is pumped from the bottom of the vessel to the inlet where it is distributed through spray nozzles over the surface of the drug. The movement of the solvent reduces boundary layers and the uniform distribution minimizes local concentrations, leading to more efficient extraction in a shorter time. Like the normal maceration, however, extraction is incomplete, since mass transfer will cease when equilibrium is set up. This problem can be overcome by using a counter current multistage extraction process.

Procedure : For Multi Stage Extractors

1. Fill extractor with drug, add solvent and circulate. Run off to receiver 1.
2. Refill extractor with solvent and circulate. Run off to receiver 2.
3. Refill extractor with solvent and circulate. Run off to receiver 3.
4. Remove drug from extractor and re-charge. Return solution from 1 to extractor. Remove for evaporation.
5. Return solution from 2 to extractor and circulate. Run off to receiver 1.
6. Return solution from 3 to extractor and circulate. Run off to receiver 2.

7. Add fresh solvent to extractor and circulate. Run off to receiver 3.
8. Remove drug from extractor and re-charge. Repeat the cycle.

For Single Stage Extractor

1. Place suitable solid material (either crushed or cut material– minced for sufficient time to diffuse the menstruum through the cell wall) in a closed vessel and add the whole chosen menstruum.
2. The system is stand for seven days with occasional shaking.
3. The system is then strained off and the solid residue (marc) is pressed.
4. The strained and the expressed liquids are united and clarified by subsidence or filtration.

Diagram :

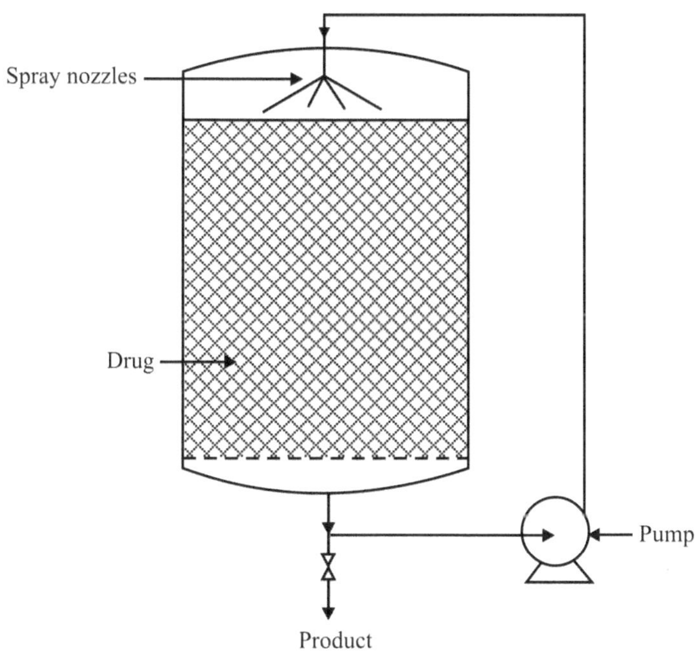

Fig. 23.1 Circulatory Extraction.

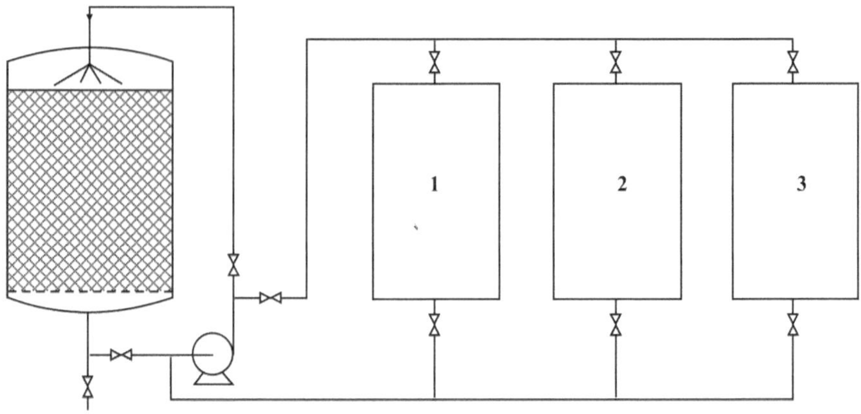

Fig. 23.2 Multiple Stage Extraction.

Report :

Bibliography :

1. Introduction to chemical engineering, 7th Edition, by **Walter L. Badger and Juliust T. Banchero** page no.348 – 349.
2. Pharmaceutical Engineering, 1st Edition 2003, by **K. Sambamurthy,** page no. 175 – 176.
3. Unit operations of chemical engineering, 6th Edition, by **M.C. Cabe, Smith, harriott,** page no. 739 – 742.
4. Cooper and Gun's, Tutorial Pharmacy, 6th Edition, Edited by **S. J. Carter,** page No. 255 – 256.

Experiment No : 24 **Dt :**

Determination of Solubility Curve

Aim : To determine the effect of temperature on solubility of NaCl by constructing the Solubility Curve.

Requirements : NaCl, 0.1N $AgNO_3$, boiling tubes, Ammonium thiocyanate, thermometer, burette, pipette, conical flask, water bath, etc.

Principle : Solubility can be defined as the concentration of solute in a saturated solution at certain temperature. (or) In quantitative way, the spontaneous interaction of two or more substances and form a homogeneous molecular dispersion is called *Solubility*. The solubility of solute in a solvent increase with the increase in temperature; if the solute absorbs heat when it dissolves i.e., it is an endothermic process. In this case the heat of solution is possibly converted, if the process is exothermic, i.e. if heat is evolved. Then solubility decreases with increase in temperature. One of the methods involves the preparation of a saturated solution of a solute at different temperature and determining the concentration in a saturated solution. Then the graph is plotted with temperature on X-axis & concentration on Y-axis to obtain solubility curve. The curve obtained is flat; the increasing temperature is very little.

Solubility curves are useful in predicting the experimental conditions desired for crystallizing a substance. Since super saturation is achieved by reducing the temperature, the influence of temperature on the solubility of a substance is important.

Procedure :

1. Small quantity of NaCl is taken in a conical flask containing definite volume of distilled water and dissolves it by stirring.
2. Add a little excess of solid till the saturated solution becomes super saturated.
3. Prepare 0.1N $AgNO_3$ solution and standardize using same strength of Ammonium thiocyanate solution.

4. Pipette out 5 ml of supersaturated solution using cotton plug and estimate the amount of NaCl present in 1 ml of supersaturated solution.
5. Calculate the NaCl solubility in 100 ml of water.
6. Similarly take 5 ml water in another boiling tube keep in the water bath and heat up to 40 °C.
7. Prepare supersaturated solution of NaCl at 40 °C.
8. Pipette out 5 ml of supersaturated solution using cotton plug and estimate the amount of NaCl present in 1 ml of supersaturated solution.
9. Repeat the same procedure at 50 °C, 60 °C, and 70 °C.
10. Plot the graph between Temperature on X-axis and Weight on Y-axis, and give inference.

Observation :

Estimation of NaCl :

Take the sample and add 50 ml of 0.1N $AgNO_3$. To this add 3 ml of HNO_3, 5 ml of nitrobenzene and 2 ml of ferric ammonium sulphate solution. Shake well and titrate with 0.1N ammonium thiocyanate until colour becomes reddish yellow.

Each ml of 0.1N $AgNO_3$ = 0.005845 gm of NaCl.

%w/w of NaCl =

$$\frac{\text{Volume of } AgNO_3 \times \text{(blank-exp) Normality of } AgNO_3 \times 0.005845}{\text{Weight of Sample} \times 0.1} \times 100$$

Table 24.1

S. No.	Temperature (° C)	Volume of burette	Amount of NaCl in 5 ml filtrate
1	Room temperature		
2	40		
3	50		
4	60		
5	70		

Report :

Bibliography :

1. Introduction to chemical engineering, 7th Edition, by **Walter L. Badger and Juliust T. Banchero** page no.142 – 143.
2. Pharmaceutical Engineering, 1st Edition 2003, by **K. Sambamurthy,** page no. 74 – 75.
3. Unit operations of chemical engineering, 6th Edition, by **M.C. Cabe, Smith, harriott,** page no. 315 – 319.
4. Pharmaceutical Engineering principles and practices, 6th Edition, by **C.V.S. Subrahmanyam**, page No. 124 – 125.

Experiment No : 25 **Dt :**

Fluid Bed Dryer (Demonstration)

Aim : To perform the drying of granules by using fluid bed drier.

Requirements : Fluidized bed dryer, granules, etc

Principle : In the fluidized bed dryers, a fan mounted in the upper part of the apparatus induces the fluidizing air stream. This is heated to the required temperature in an air heater and flows upward through the wet material, which is contained in a drying chamber fitted with a wire mesh support at the bottom. (Hot air is passed at high pressure through a perforated bottom of the container containing granules to be dried) The airflow rate is adjusted by means of a damper, and a bag (collector) filter is provided at the top of the fine particles (the granules are lifted from the bottom and suspended in the air. This condition is called *fluidized state*). The unit is called **batch-type dryer**. Dryer capacities range from 5 kg to 200 kg and the average drying time is about 20 to 40 min. Because of the short drying time and excellent mixing action of the dryer, no hot spots are produced, and higher drying temperatures can be employed than are used in conventional tray and truck dryers.

Procedure :

1. Prepared wet granules are placed in the detachable bowl for drying.
2. The bowl is pushed into the dryer.
3. Air is allowed to pass through a pre-filter, which gets heated by passing through heaters.
4. The hot air flows through the bottom of the bowl simultaneously fan is allowed to rotate. The air velocity is gradually increased.
5. When the velocity of the air is greater than settling velocity of granules, the granules remain partially suspended in the gas stream.
6. After some time, a point of pressure is reached at which frictional drag on the particles is equal to the force of gravity.
7. The granules rise in the container because of high velocity of air (1.5 to 7.5 m/min).

8. The air surrounds each and every granule to completely dry them.
9. The air leaves the dryer by passing through the bag filter. The particles remain adhere to the inside surface of the bags.
10. The bags are shaken to remove the particles.
11. The residence time for drying is about 40 min. After some time bowl is taken out for discharge.

Applications :

1. Drying of granules in the production of tablets.
2. Drying of granules in the production of capsules.

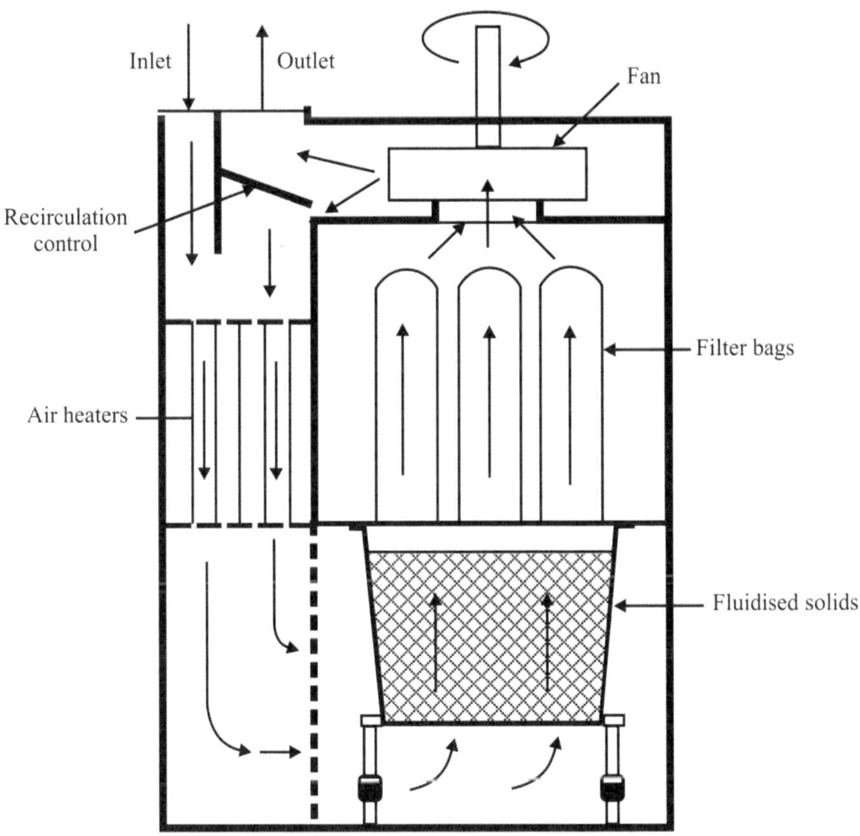

Fig. 25.1 Fluidized Bed Dryer.

Report : Using fluidized bed dryer, the dried product (granules) obtained.

Bibliography :

1. Pharmaceutical Engineering, 1st Edition 2003, by **K. Sambamurthy,** page no. 309 – 312.
2. Unit operations of chemical engineering, 6th Edition, by **M.C. Cabe, Smith, harriott,** page no. 171 – 180.
3. Pharmaceutical Engineering principles and practices, 6th Edition, by **C.V.S. Subrahmanyam**, page No. 397 – 399.
4. The theory and practice of Industrial Pharmacy, 3rd Edition, by **Leon Lachman, Herbert A. Lieberman,** page no. 58 – 60.
5. Cooper and Gun's, Tutorial Pharmacy, 6th Edition, Edited by **S. J. Carter,** page No. 276 – 277.

Experiment No : 26 **Dt :**

Filter Press (Demonstration)

Aim : To perform the filtration by using filter-press (Plate and frame filter press).

Requirements : Plate of chamber filter press, filter cloth, Movable head, etc.

Principle : Filtration may be defined as a process of separation of solids from a fluid by passing the same through a porous medium that retains the solids, but allows the fluid to pass though.

The mechanism involved in plate and frame filter press is *surface filtration.* The slurry enters the frame by pressure and flows through the filter medium. The filtrate is collected on the plates and sent to the outlet. A number of frames and plates are used so that surface area increases and consequently large volumes of slurry can be processed simultaneously with or without washing.

Theory : The oldest standard construction of a filter is a filter press. The principal subdivisions are chamber presses and plate-and-frame presses. The filter press is the most versatile filters since the number and type of filter sheets can be varied to suit a particular requirement. It can be used for coarse to fine filtrations, and by special conduit arrangements for multistage filtration within a single press. Filter press can also be used for sterilization of liquids. The filter press is the most economical filter per unit of filtering surface, and material of construction can be chosen to suit any process conditions. The normal range of flow is three gallons per minute per square foot of filter surface at pressures of up to 25 psi.

The plate and frame (also called *flush–plate*) filter press is the simplest of all pressure filters and is the most widely used filter. Filter presses are used for a high degree of clarification of the fluid and for the harvesting of the cake.

Construction : The simplest and cheapest type is the Recessed plate press/Camber press. A heavy fixed head plate of cast iron is mounted on a suitable frame and has the necessary pipe connections on it. Extending from the head plate are two horizontal bars supported at the other end by the end frame. These bars are support the plates of the press by lugs A. the press plates are usually of cast iron, 1 to 3 ft. wide, ½ in. thick and with a raised edge from ½ to 2 in. around the outside. The plates may be either square or round. In the center of each press plate is a hole B, which is to be in line with a connection on the head of the press where the feed is introduced. Over each press plate is thrown a sheet of filter cloth with a hole cut in the center to register with the hole in the plate. When all the plates have been so dressed, a heavy follower plate is placed behind the whole assembly and the plates are pressed tightly together by means of a heavy screw or a hydraulic pressure device. The cloth serves as a gasket between the edges of the adjacent plates.

Procedure :

1. Now the material to be filtered is pumped through the connection at the center of the head of the press to fill the openings between the cloths.
2. As material continues to be pumped in, the escapes through the holes C cored in the bottom of each plate, connecting with an external outlet D.
3. This external outlet usually discharges into as open launder.
4. Collector should be connecting to the discharge pipe. Collect the filtrate that is filtered.

Report : By using filter press, the filtrate has collected.

Bibliography :

1. Introduction to chemical engineering, 7th Edition, by **Walter L. Badger and Juliust T. Banchero** page no.556 – 557.
2. Pharmaceutical Engineering, 2003 1st Edition, by **K. Sambamurthy,** page no. 253 – 254.
3. Unit operations of chemical engineering, 6th Edition, by **M.C. Cabe, Smith, harriott,** page no. 993 – 994.
4. The theory and practice of Industrial Pharmacy, 3rd Edition, by **Leon Lachman, Herbert A. Lieberman,** page no. 159 – 160.

Experiment No : 27 **Dt :**

Evaporation Under Reduced Pressure (Roto Evaporator–Demonstration)

Aim : To perform the evaporation under reduced pressure by using Roto evaporator.

Requirements : Roto evaporator, dye in methanol, thermostat.

Principle : In this method, the evaporation is carried under reduced pressure, due to this; the mixture is boiled at the below normal boiling point of the mixture. Usually the temperature is maintained at 60 °C. Due to rotation of flask the end product is obtained as thin film around the walls of the flask.

Figure :

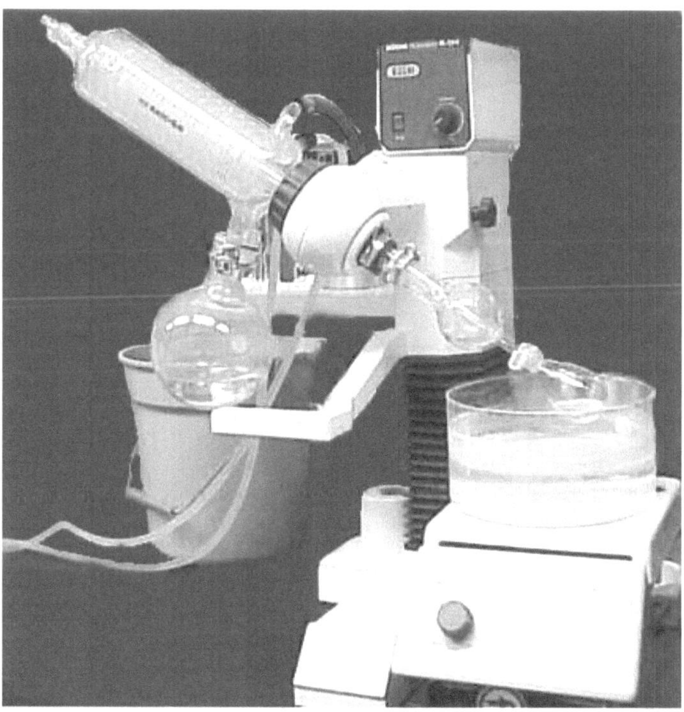

Procedure :

1. Roto evaporator consists of a flask and a special condenser with provision for inlet and outlet water and vacuum as shown in the figure.
2. The flask is rotated by means of a motor.
3. The dye in methanol is taken in a flask and connected to condenser, then immersed into water bath (temperature maintained at 60 °C).
4. Connect the inlet, and outlet water pipes. Vacuum is given to condenser.
5. Then switch on the motor and adjust the speed of rotation.
6. The solvent vaporizes, enters into condenser then get condensed and is collected in receiver.
7. The end product is obtained as thin film around the walls of the flask.

Report : Using Roto evaporator, which is carried out order reduced pressure, the dry product is obtained in the flask.

Experiment No : 28 **Dt :**

Extraction of Organised Substances by Percolation Method

Aim : To determine the extraction of organized substances by using Percolation method.

Requirements : Percolator, Cotton plug, Glass beads, Drug seeds, fine Sand, Filter Paper.

Theory:

Percolation : The term 'percolation' is given to that method of filtering in which the oil to be decolorized is passed through a bed of granular adsorbent aided via the help of gravity.

1. ***Comminution.*** The principles of size reduction are covered in experiment 06. Remember if the particles are too fine a solid cake may occur, this will affect the downward flow of menstruum and will most certainly lead to the formation of 'dry pockets' within the body of the material, which will escape extraction. If the material is too coarse then interstices are formed through which there is a speedy percolation of menstruum which produces an incomplete extraction and will require excessive volumes of menstruum to exhaust the marc.
2. ***Imbibition***. The word is derived from the Latin meaning 'to drink in'. The comminuted drug thoroughly moistened with a portion of the menstruum. This is best done in a lidded container of a suitable size. The moistened drug is allowed to stand for a period of four hours to allow the drug to imbibe the menstruum and thereby swell to its maximum capacity. The container used should be large enough to accommodate the expansion of the drug.
3. ***Packing.*** On completion of imbibition the drug should be passed through a number #10 sieve to break up any lumps that may have

formed. The drug is then transferred to the percolation vessel in portions. Each portion should be firmly packed but not so firmly that liquid is forced from the drug but sufficient to exclude any air pockets.

4. ***Maceration.*** Open the stopclock on the bottom of the percolator and pour sufficient quantity of menstruum in portions and allow percolating through the packed drug. If the menstruum drips through the stopclock in less than 10 minutes, the drug is too loosely packed. If the first drop takes 25 minutes or more then the drug is too tightly packed. If all is well, then close the stopclock and pour in sufficient menstruum to leave a layer 1 or 2 cm deep over the drug. Cover the percolator and leave to macerate for 48 hours in a warm dark place at a temperature not exceeding 25°C.

Procedure :

In this process the raw material (Ex: an organized vegetable substance-drug) in a suitably powdered form, is packed into percolator and the solvent is allowed to percolate through it. It may be considered that a solvent is single component like *n*-hexane while the term menstruum covers a solvent mixture like 80% alcohol in water. If the dry material is packed into the percolator, it will swell (increase in volume) when the solvent is added (because of adsorption of the solvent), the extent of swelling increasing with increasing aqueous nature of the solvent. This swelling reduces or blocks the flow of the solvent, thus particles may be washed down the column to settle at lower levels reducing the porosity drastically, resulting in the blocking of the column and making the column non-uniform. The finer particles may even be washed out of the column. These difficulties can be prevented by preliminary uniform moistening of the raw material with the menstruum for a period of four (4) hours in a separate closed vessel, a process called *imbibition*. During this period, the drug is allowed to swell to maximum extent, if require add more amount of menstruum.

The solvent is penetrating into the drug molecule thereby enabling the material to be more evenly packed and allowing menstruum to flow more uniformly even the wet material packed more uniformly than dry powder.

After imbibition, the drug is packed evenly into the percolator, as showed in the diagram given below.

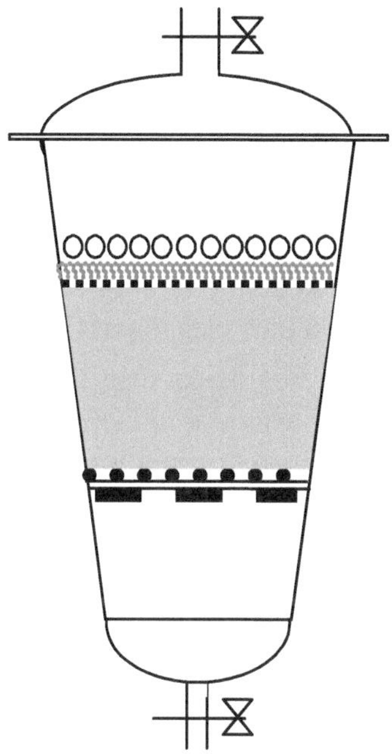

Percolation Diagram.

Diagram shows a type of percolator used on the Lab scale/Industrial scale.

1. The drug is supported on a perforated metal plate covered with a layer of sacking or straw.
2. The percolator is provided with a removable lid, which contains portholes for packing the drug, for running in the solvent and for observing the flow of solvent.
3. The outlet from the percolator is fitted with a tap and pipe line to remove the percolate for subsequent processing or to use it as a menstruum for the second percolator in series for more efficient use of the menstruum by carrying out the extraction in a countercurrent manner.
4. The imbibed drug is packed over a loose plug of tow or other suitable material previously moistened with the solvent, material construction should be as layer by layer and pressing it with a suitable implement to give even compression.

5. After packing is over, a piece of filter paper is placed on the surface followed by a layer of clean sand such that the top layers of the drug are not disturbed when solvent is admitted for extraction. A layer of glass beads poured on the top layer for pressure to produce on the drug material.
6. Sufficient menstruum is now poured over the drug slowly and evenly to saturate it, keeping the tap at the bottom to allow the occluded gases between particles to pass out.
7. When the menstruum begins to drip through the tap, the tap is closed; sufficient menstruum is added to maintain a small layer above the drug and allowed to stand for 24 hours.
8. The 24 hours maceration period allows the solvent to diffuse through the drug, solubilise the constituents and leach out the soluble material.
9. After the maceration, the outlet is opened and the solvent is percolated at a controlled rate with continuous addition of fresh solvent.
10. In general 75% of the volume of the finished product is collected, the marc is pressed and expressed liquid is added to the percolate giving about 80-90% of the final volume. After assay, the volume is adjusted with calculated quantities of fresh menstruum.

Applications :

1. Self-preservative for preventing microbial contamination.
2. It is best method for extracting vegetable sources published by *British Pharmacopoeia* & *British Pharmaceutical codex.*
3. Percolation method mainly depends on flow of the solvent through the powdered drug, to yield greater concentration products, where as maceration methods yields less when compared to percolation method.

Bibliography :

1. Pharmaceutical Engineering by K. Sambamurthy, p.no. 176 – 178.
2. Cooper and Gunn's Tutorial Pharmacy by S.J. Carter, p.no. 252 – 255.

Experiment No : 29 **Dt :**

Continuous Hot Extraction Method

Aim : To determine the extraction of drug substances by using soxholation method.

Requirements : Soxhlet apparatus, Cotton plug, Drug seeds, Filter Paper, solvent (suitable).

Parts of Soxhlet apparatus:

1. Stirrer bar / anti-bumping granules
2. Still pot (extraction pot) - still pot should not be overfilled and the volume of solvent in the still pot should be 3 to 4 times the volume of the soxhlet chamber.
3. Distillation path
4. Soxhlet Thimble
5. Extraction solid (residue solid)
6. Syphon arm inlet
7. Syphon arm outlet
8. Expansion adapter
9. Condenser
10. Cooling water in
11. Cooling water out

Here showing a photograph taken for actual diagram setup:

A **Soxhlet extractor** is a piece of laboratory apparatus invented in 1879 by Franz von Soxhlet.[1] It was originally designed for the extraction of a lipid from a solid material. However, a Soxhlet extractor is not limited to the extraction of lipids.

Typically, a Soxhlet extraction is only required where the desired compound has a limited solubility in a solvent, and the impurity is insoluble in that solvent. If the desired compound has a high solubility in a solvent then a simple filtration can be used to separate the compound from the insoluble substance.

Fruit extraction in progress. The sample is placed in the thimble.

Normally a solid material containing some of the desired compound is placed inside a thimble made from thick filter paper, which is loaded into the main chamber of the Soxhlet extractor. The Soxhlet extractor is placed onto a flask containing the extraction solvent. The Soxhlet is then equipped with a condenser.

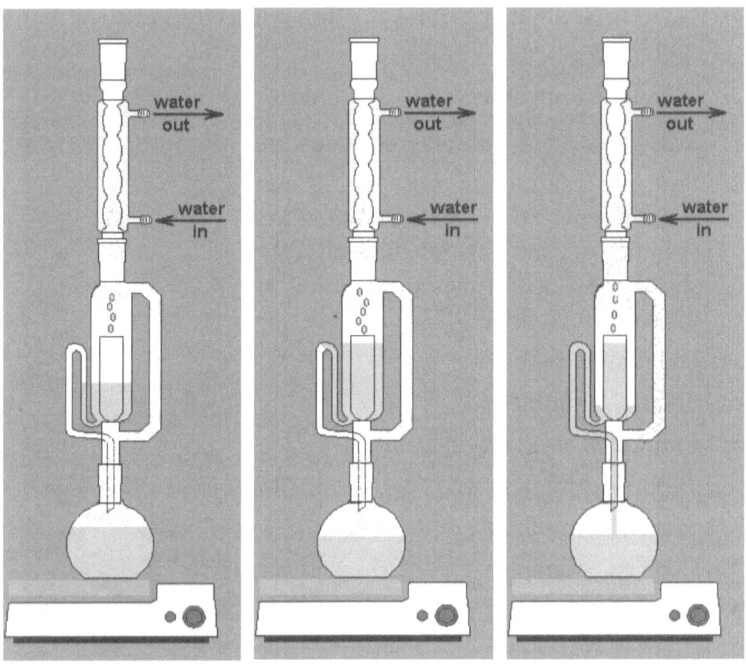

The solvent is heated to reflux. The solvent vapour travels up a distillation arm, and floods into the chamber housing the thimble of solid. The condenser ensures that any solvent vapour cools, and drips back down into the chamber housing the solid material.

The chamber containing the solid material slowly fills with warm solvent. Some of the desired compound will then dissolve in the warm solvent. When the Soxhlet chamber is almost full, the chamber is automatically emptied by a siphon side arm, with the solvent running back down to the distillation flask. This cycle may be allowed to repeat many times, over hours or days.

During each cycle, a portion of the non-volatile compound dissolves in the solvent. After many cycles the desired compound is concentrated in the distillation flask. The advantage of this system is that instead of many portions of warm solvent being passed through the sample, just one batch of solvent is recycled.

After extraction the solvent is removed, typically by means of a rotary evaporator, yielding the extracted compound. The non-soluble portion of the extracted solid remains in the thimble, and is usually discarded.

Limitations of this Process are;

1. It is not useful when the raw material contains thermolabile active, the composition of the vapour will be different from the liquid constituents because not only the extraction is carried out at an elevated temperature but the extract in the flask is maintained in the hot condition until the process is complete.
2. It can be used only with pure solvents or with solvent mixtures forming azeotropes, if an ordinary binary mixture is used as the menstruum composition.

Bibliography :

1. Soxhlet, F. Die gewichtsanalytische Bestimmung des Milchfettes, Polytechnisches **J. (Dingler's)** 1879, 232, 461.
2. *The Origin of the Soxhlet Extractor* **William B. Jensen** Vol. 84 No. 12 December 2007, Journal of Chemical Education 1913.
3. Cooper and Gun's, Tutorial Pharmacy by **S.J.Carter**, page number 257 – 259.
4. Pharmaceutical Engineering by **K. Sambamurthy**, P.No. 187 – 188.

Experiment No : 30 **Dt :**

Factors Affecting the Rate of Evaporation

Aim : To determine the effect of factors influencing the rate of evaporation

Requirements : Water, glycerin, sugar syrup (I.P), NaCl, beakers of different sizes, glass rods, heat source.

Theory : Theoretically Evaporation means, simply vaporization from the surface of a liquid. Practically evaporation means, the removal of liquid from a solution by boiling the liquid in a suitable vessel and withdrawing the vapour, leaving a concentrated liquid residue.

This means that heat will be necessary to provide the latent heat of vaporization and generally the rate of evaporation is concentrated by the rate of heat transfer. In all the following methods below; thermal heating & time is considered to be a pre-requisite for evaporation, because evaporation rate depends on temperature & time. If time and temperature increases rate of evaporation increase.

Evaporation is a process of vaporizing large quantities of volatile liquids, solvents, water, gives a concentrated product.

Principle : Evaporation is a surface phenomenon, i.e. mass transfer takes place from the surface. The rate of evaporation depends on several factors such as temperature, temperature & time of evaporation, temperature & moisture content, viscosity, vapor pressure, surface area, concentration of the solution to be evaporated, type of the product required & lastly economic factors also influence the rate of evaporation.

1. Temperature – If the temperature increase, the rate of evaporation increases, so rate of evaporation is directly proportionate to the temperature.
2. Time – If the time increase, the rate of evaporation increases, so rate of evaporation is directly proportionate to the time.
3. Viscosity – If the viscosity increase, the rate of evaporation decreases, so rate of evaporation is indirectly proportionate to the viscosity.

4. Concentration – If the concentration increase, the rate of evaporation decreases, so rate of evaporation is indirectly proportionate to the concentration.
5. Moisture – If Moisture content is more, the rate of evaporation increases with slow rate, if moisture content is less at the end point, rate of evaporation is rapid, so rate of evaporation is directly proportionate to the concentration.

Practically rate of evaporation is calculated based on below mentioned formula;

$$\text{Rate of Evaporation} = \frac{W_1 - W_2(g)}{\text{Timc (min)}}$$

W_1 = Initial weight of solution

W_2 = Final weight of solution

$(W_1 - W_2)$ = Quantity of solution evaporated

Procedure :

1. ***To determine the effect of Surface area on Rate of Evaporation :*** Take accurately 100 g of water into 3 different size beakers (100, 250, 500 ml). Determine their surface areas by measuring their radius, Surface area = πr^2. Heat the beakers by keeping on heating pan at 80 °C for 30 min, then weigh the beaker weight and note it down, then continue this process for remaining beakers and calculate the rate of evaporation as per given above formula. Plot a graph by taking Rate of Evaporation on Y-axis and Surface area on X-axis. Report the results and conclude with the standard results.
2. ***To determine the effect of Viscosity on Rate of Evaporation :*** Take accurately 25 g of water, Sugar syrup 66.67% w/w (I.P.), glycerine: water ratio (1:1) into 3 different 100 ml capacity beakers. Determine their viscosities, and heat the beakers by keeping on heating pan at 80 °C for 30min, then weigh the beaker weight and note it down, then continue this process for remaining beakers and calculate the rate of evaporation as per given above formula. Plot a graph by taking Rate of Evaporation on Y-axis and Viscosity on X-axis. Report the results and conclude with the standard results.
3. ***To determine the effect of Concentration on Rate of Evaporation :*** Prepare salt solution of NaCl with different concentrations like 10, 20, 30% w/v and take weighed quantity of 50g solution into 3 different 100 ml capacity beakers, and heat the beakers by keeping on heating pan at 80 °C for 30 min, then weigh the beaker weight and note it

down, then continue this process for remaining concentration solution beakers and calculate the rate of evaporation as per given above formula. Plot a graph by taking Rate of Evaporation on Y-axis and Concentration on X-axis. Report the results and conclude with the standard results.

4. ***Effect of Impurities in solution :*** Rate of evaporation can be either increased or decreased based on the nature of the impurities present in the solution.

5. ***Effect of Time on rate of evaporation :*** If time increases the rate of evaporation increases with all the parameters, which is shown in figure as example. Because time of exposure is more to the temperature, so evaporation will be more, and then rate of evaporation will more.

Applications :

1. Mainly pharmaceutical inorganic materials are prepared.
2. Evaporation employed in manufacture of galanicals (extraction preparations like insulin, penicillin's, etc.)
3. This evaporation is commonly used in unit operation in pharmaceutical industries and in bulk drug manufacturing area.
4. There are many applications in other fields other than pharmaceutical field.

Observations :

To determine the effect of Surface area on rate of evaporation:

S. no.	Surface area of beaker	Initial weight (w_1 g)	Final weight (w_2 g)	Remaining weight (w_1-w_2)	Rate of evaporation
1					
2					
3					

To determine the effect of Viscosity on rate of evaporation:

S. no.	Viscosity	Solution	Initial weight (w_1 g)	Final weight (w_2 g)	Remaining weight (w_1-w_2)	Rate of evaporation
1						
2						
3						

To determine the effect of Concentration on rate of evaporation:

S. no.	Concentration of solution	Initial weight (w_1 g)	Final weight (w_2 g)	Remaining weight (w_1-w_2)	Rate of evaporation
1					
2					
3					

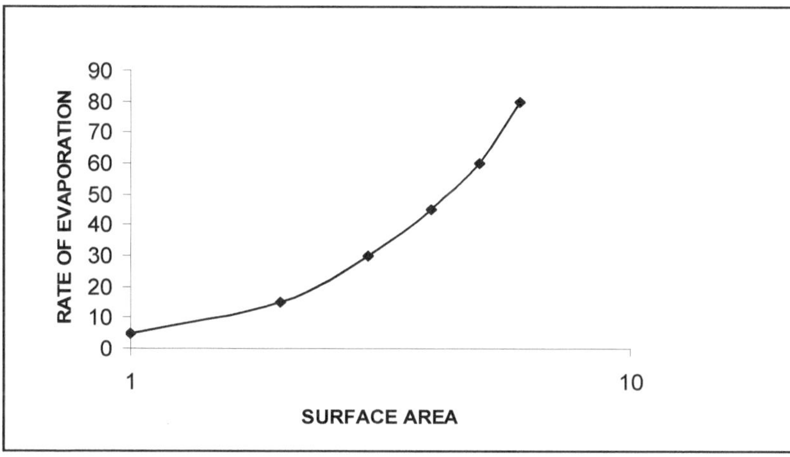

Fig. 30.1 Effect of Surface Area on Rate of Evaporation.

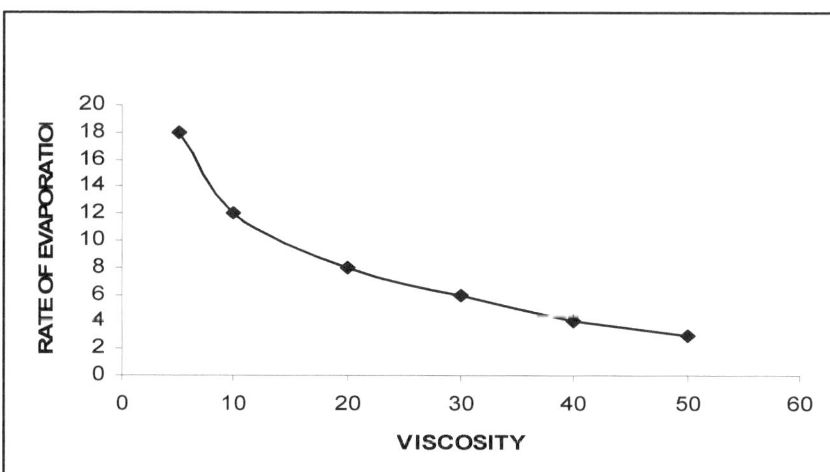

Fig. 30.2 Effect of Viscosity on Rate of Evaporation.

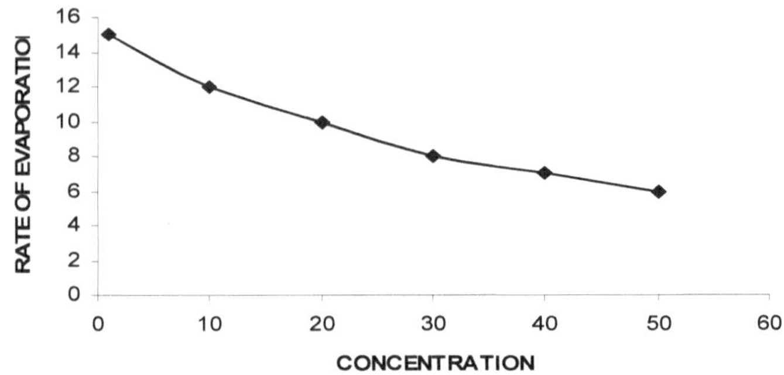

Fig. 30.3 Effect of Concentration on Rate of Evaporation

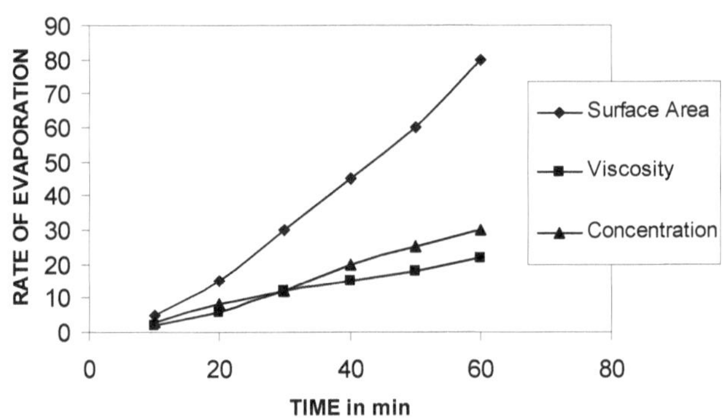

Fig. 30.4 Factors influencing the rate of Evaporation.

Bibliography :

1. Introduction to chemical engineering, by **Badger & Banker**; p.no. 170 – 178.
2. Cooper & Gunn's Tutorial Pharmacy by **S.J.Carter**; p.no. 164 – 173.
3. Pharmaceutical Engineering Principles & Practices by **C.V.S. Subrahmanyam** p.no. 337 – 340.
4. Pharmaceutical Engineering by **K. Sambamurthy** p.no. 89 – 101.

Experiment No : 31 **Dt :**

Effect of Particle Size on Sedimentation Rate

Aim : To estimate the sedimentation rate by settling method for the formulation of calamine lotion.

Material required : Calamine lotion, measuring glass cylinders, etc.

Principle : The process of settling is best described by batch-settling tests in glass cylinders. The size and type of particles to be removed have a significant effect on the operation of the sedimentation tank. Because of their density, sand or silt can be removed very easily. The velocity of the water-flow channel can be slowed to less than one foot per second, and simple gravitational forces will remove most of the gravel and grit. In contrast, colloidal material, small particles that stay in suspension and make the water seem cloudy, will not settle until the material is coagulated and flocculated by the addition of a chemical, such as an iron salt or aluminum sulfate.

The shape of the particle also affects its settling characteristics. A round particle, for example, will settle much more readily than a particle that has ragged or irregular edges.

All particles tend to have a slight electrical charge. Particles with the same charge tend to repel each other. This repelling action keeps the particles from congregating into floccules and settling.

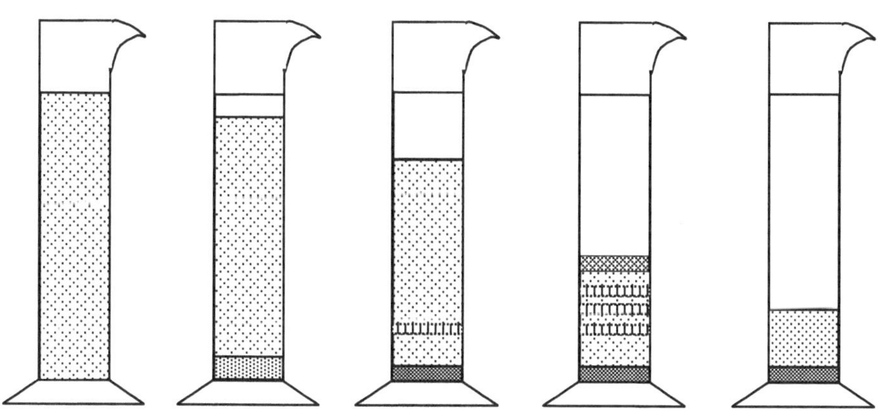

Procedure :

1. Prepare calamine lotion by using suspending agent bentonite as per specified formula in the textbooks.
2. Transfer the 100 ml sample into the measuring cylinder and allow it to settle down with out any disturbance.
3. Measure the initial volume of the sediment settled down, and note it down.
4. After 15 min, check the volume of sediment settled down and note it down.
5. Repeat the above step for every 15 min until 2 hr.
6. Plot a graph between **volume of sediment** on Y-axis and **Time** on X-axis.

Table 31.1 Preparation of calamine lotion

Ingredients	**For 100 ml**
Calamine	15 g
ZnO	5 g
Bentonite	3 g
Sodium citrate	0.5 g
Liquified phenol	0.5 ml
Glycerin	1 ml
Water	Upto 100 ml

Graph : Refer Experiment No. 18, Titled Sedimentation by centrifugation.

References :

1. Introduction to chemical Engineering by **Walter L Bagder and Juliust T. banchero** page. no. 593 – 599.
2. Pharmaceutical Engineering by **K. Sambamurthy**, page. no. 282 – 286.
3. Unit operations of chemical engineering by **M.C. Cabe, Smith harriott**, page.no.1000-1003.
4. Physical Pharmacy, by **Alfred Martin**, page no. 547 – 548.

Definitions

Drying

Drying : Drying is defined as the removal of small amounts of water or other liquid from a material by the application of heat.

Bound water : Bound water is the minimum water held by the material that exerts an equilibrium vapour pressure less than the pure water at the same temperature.

Unbound water : unbound water is the amount of water held by the material that exerts an equilibrium vapour presser equal to that of pure water at the same temperature.

Equilibrium moisture content (EMC) : It is the amount of water present in the solid that exerts a vapour pressure equal to the vapour pressure of the atmosphere surrounding it.

Hygroscopic Substances : Substances containing bound water are often called hygroscopic substances.

Non-hygroscopic Substances : Substances containing unbound water are often called non-hygroscopic substances.

Desorption : When air is continuously passed over the solid containing moisture more than EMC, then solid losses water continuously till EMC is reached. This phenomenon is known as desorption.

Sorption : When air is continuously passed over the solid containing moisture less than EMC, then solid adsorb water continuously till EMC is reached. This phenomenon is known as sorption.

Free moisture content (FMC) : Free moisture content (FMC) is the amount of water that is free to evaporate from the solid surface.

Drying rate : The ratio of weight of water (kg) in sample by time into weight of the dry solid.

% Loss on drying : The % ratio of mass of water in sample (kg) by total mass of the wet sample (kg).

% Moisture Content (MC) : The % ratio of mass of water in sample (kg) by mass of the dry sample (kg).

Drying rate curve : A graph is plotted between FMC on X-axis and drying rate on Y-axis, the curve so obtained is called drying rate curve.

Critical Moisture Content : The rate of diffusion is equal to the rate of evaporation. The moisture content at the end of constant rate is referred to as the critical moisture content.

Eutectic Point : The pressure and temperature at which the frozen solid vaporizes without conversion to a liquid is referred to as the eutectic point.

Mier's Theory : Under ideal conditions of crystallization nucleus formation starts at super solubility curve and crystal growth begins.

Crystallization

Crystallization : crystallization is the spontaneous arrangement of the particles into a repetitive orderly array i.e., regular geometric patterns.

Sublimation : Crystallization can takes place directly from vapour of a substance, ex: Solid camphor from camphor vapour, solid iodine from iodine vapour such process is known as *sublimation.*

Crystal : A crystal can be defined as a solid particle, which is formed by the solidification process in which structural units are arranged, by a fixed geometric pattern or lattice.

Crystal lattice : It is defined as an orderly internal arrangement of particles in three-dimensional spaces.

Space lattice : The three dimensional arrangement of particles in a crystal is also known as space lattice.

Unit cell : The smallest geometric portion, which repeats to build up the whole crystal, is called a Unit cell.

Faces : A crystal is bounded by plane surfaces called faces.

Axial Angle : In the crystal, the angle between the two perpendiculars to the intersecting faces is termed as the axial angle.

Axial Length : Axial length can be defined as the distance between the centers of two atoms.

Crystal system (or) forms : A finite number of symmetrical arrangements are possible for a crystal lattice and these may be termed as crystal forms or systems.

Crystal Habit : The term crystal habit is used to denote the relative development of the different types of faces and not to the shape of the resulting crystals.

Crystalline Solids : Crystals that have definite shape and an orderly arrangement of units of incompressible is called as crystalline solids.

Amorphous Solids : Amorphous solid do not have specific shape and their structural units are arranged randomly in the solid.

Polymorphs : Certain drugs can exist in more than one crystalline form. Such a phenomenon is known as polymorphism.

Drug Hydrates : Some drugs have greater tendency to associate with water. The resulting substance is referred to as drug hydrates.

Pseudomorphs : Certain drugs have greater tendency to associate with solvents to produce crystalline forms of solvates. These solvates are also known as pseudomorphs.

Isomorphism : When two or more substances possess the same crystalline form, the crystals of one such substance can be grown in the saturated solution of the other. This phenomenon is known as isomorphism.

Nucleation : Nucleation refers to the birth of very small bodies of a new phase within a homogenous supersaturated liquid phase.

Crystal Growth : Crystal growth is a diffusion process and surface phenomenon, in which solute molecules or ions reach the faces of a crystal by diffusion, and organized into space lattice.

Meta-stable state : The region enclosed between the two curves called normal solubility curve AB and super solubility curve FG is referred as meta-stable state.

Solubility curve : The graph drawn by taking temperature on X-axis and solubility on Y-axis gives the solubility curve.

Caking of crystals : Caking can be defined as the process of formation of clumps or cakes when crystals are improperly stored.

Critical Humidity : Critical humidity is the humidity above which crystals absorb moisture and below which they do not absorb moisture.

Cubic : Three equal axes at right angle to each other.

Tetragonal : Three axis all right angles, one longer than the other two.

Orthorhombic : Three axes all at right angle, but all of different lengths.

Hexagonal : Three equal axes in one plane at 60° to each other, one at right angle to this plane but not necessarily the same length as the other.

Monoclinic : Two axes at right angles in one plane, and a third axis at some odd angle to this plane.

Triclinic : Three axes at odd angles to each other.

Distillation

Distillation : Distillation is defined as the separation of the components of a liquid mixture by a process involving vaporization and subsequent condensation at another place.

Simple Distillation : It is a process of converting a single constituent from a liquid (mixture) into its vapour transferring the vapour to another place & recovering the liquid by condensing the vapour, usually by allowing it to come in contact with cold surface. This is also known as **Differential distillation.**

Flash Distillation : It is defined as a process in which the entire liquid mixture is suddenly vapourized (flash) by passing the feed from a high-pressure zone to a low pressure. It is also known as **Equilibrium distillation.**

Fractional distillation : It is a process in which vapourization of liquid mixture gives rise to a mixture of constituents from which the desired one is separated in pure form. This is also known as **Rectification.**

Reflux ratio : It is the quotient of the amount of liquid returning through the column to the amount collected into the receiver during the same interval of time.

Azeotropic Distillation : It is a method in which azeotropic mixture is broken by the addition of a third substance, which forms a new azeotrope with one of the components.

Extractive Distillation : It is a distillation method in which the third substance added to azeotropic mixture is relatively non-volatile liquid compared to the components to be separated.

Distillation Under reduced Pressure : It is defined as distillation process in which the liquid is distilled at a temperature lower than its boiling point by the application of vacuum.

Steam Distillation : It is a method of distillation carried with the aid of steam and is used for the separation of high boiling substances from non-volatile impurities.

Molecular Distillation : It is defined as a distillation process in which each molecule in the vapour phase travels mean free path and it's condensed individually without intermolecular collisions on application of vacuum.

Mean free path : It is defined as the average distance through which a molecule can move without coming into collisions with another.

Destructive Distillation : It is a distillation method in which the distillate is decomposition products of the constituents of the organic matter burnt in the absence of air. This process is also known as **Dry distillation.**

Distilland : The feed liquid is known as Distilland.

Condensate : The condensed liquid is known as condensate.

Binary mixture of liquids : When two liquids are mixed together, they may be miscible with each other in all proportions. Such miscible liquids are known as binary mixture of liquid.

Raoult's Law : Raoult's law stated that the partial vapour pressure of each volatile constituent is equal to the vapour pressure of the pure constituent multiplied by its mole fraction in the solution at a given temperature.

Ideal solution : Ideal solution is defined, as the one in which there is no change in the properties of the components other than dilution, when they are mixed to form a solution.

Perfect solution : Raoult's law is obeyed by only a few solutions of liquid in liquids. These solutions are known as perfect solutions.

Real solution : Real solution is defined as the one, which show varying degrees of deviation from Raoult's law.

Dalton's law : Dalton's law of partial vapour pressure states that the total pressure exerts by a mixture of ideal gases may be considered as sum of the partial vapour pressure exerted by each gas, if alone were present and occupied the total volume.

Positive deviation : The vapour pressure is greater than the sum of the partial pressures of the individual components. Such systems are said to exhibit positive deviation from raoult's law.

Negative deviation : The vapour pressure is lower than the sum of the partial pressures of the individual components. Such systems are said to exhibit negative deviation from raoult's law.

Volatility : The volatility of any substance in a solution is defined as the equilibrium partial pressure of the substance in the vapour phase divided by the mole fraction of the substance in the solution.

Relative volatility : Relative volatility is defined as a ratio of volatility of two components (component A to the volatility of component B).

Azeotrope : Many liquid mixtures cannot be separated completely into pure components by simple distillation, because the volatilities of the components are equal. Such a mixture is known as azeotrope.

Azeotropic solution : Azeotropic solution is a solution, which distils unchanged at a constant temperature (Constant boiling mixtures).

Azeotropic mixture : The system that exhibits a maximum value in the boiling point composition such a system is known as azeotropic mixture with a minimum vapour pressure or maximum boiling point.

Humidity and Aircondition

Humidity : The pounds of water vapour carried by one pound of dry air under any given set of conditions is known as Humidity.

Saturated air : The air in which the water vapour is in equilibrium with liquid water at the given conditions of temperature and pressure is called saturated air.

Saturated Humidity : The water vapour, which is present in the saturated air, is called saturated humidity.

Percentage Humidity : % humidity is the ratio of weight of water vapour carried by one pound of dry air at any temperature and pressure to the weight of water one pound of dry air could carry if saturated at same temperature and pressure, and expressing in the result on a percentage basis.

% Relative Humidity : It is the ratio of the partial pressure of water vapour in the air-water vapour mixture to the partial pressure of water vapour at saturation (Liquid water).

Humid Heat : Humid heat is a number of Btu necessary to raise the temperature of one pound of dry air plus whatever water vapour it may carry by one ^{o}F.

Enthalpy : Enthalpy of air-water vapour mixture is the enthalpy of one pound of dry air plus the enthalpy of accompanying water vapour.

Humid Volume : Humid volume is the volume in Cu. ft. occupied by one pound of dry air and its accompanying water vapour.

Saturated Volume : It is the volume in Cu. ft of one pound of dry air plus that of the water vapour necessary to saturate it.

Dew Point : It is the temperature to which a mixture of air and water vapour must be cooled in order to become saturate. It is also known as Saturation Temperature.

Dry bulb Temperature : It is a temperature of moist air, when it is measured at rest by any instrument, which is not effected by the moisture content of air or by radiation.

Wet bulb Temperature : It is the dynamic equilibrium temperature attained by a water surface when exposed to air under adiabatic conditions.

Adiabatic Saturation Temperature : When water is sprayed into a stream of gas, equilibrium is reached between gas and water. Then the temperature of gas is called adiabatic saturation temperature.

Humidity Chart : A convenient diagram, which shows the properties of mixtures of a permanent gas and a condensable vapour is called humidity chart.

Air Conditioning : It is a process of treating air, so as to control simultaneously its temperature, humidity, cleanliness and distribution to meet the requirements of the conditioned space.

Humidification : It is a Unit operation, which involves a transfer of liquid water into water vapour.

Dehumidification : The process in which the moisture is decreased (Transfer of vapour to liquid state) in the air is known as dehumidification.

Latent heat of Vapourization : It is defined as the quantity of heat required to convert a unit mass of the liquid at its boiling point from the liquid to the vapour state without a change in temperature.

Flow of Heat

Conduction : When heat flow in a body by the transfer of momentum of individual atoms or molecules without mixing, such process is known as conduction.

Convection : When heat is achieved by actual mixing of warmer portion with cooler portion of same material, this process is known as convection.

Radiation : When the heat flows through space by means of electromagnetic waves is known as radiation.

Forced convection : Mixing of fluid may be obtained by the use of a stirrer or agitator or pumping the fluid for re-circulation. Such a process in heat transfer is designated as forced convection.

Natural convection : When a fluid is heated, the current setup may cause mixing of fluid. Such a heat transfer process is known as natural convection.

Film coefficient : Film coefficient is the quantity of heat flowing through unit area of the film for unit drop in temperature.

Parallel heat flow : When the hot fluid and the cold fluid enter the apparatus from the same end, the flow is parallel to each other. This arrangement is known as parallel flow.

Counter-current heat flow : When the hot fluid is passed through one end of the apparatus while cold fluid is passed through the other end, fluids pass and by pass each other in the opposite directions. This arrangement is known as counter-current or counter flow.

Thermal radiation : Heat transfer by radiation is known as thermal radiation.

Black body : Black body is defined as a body that radiates maximum possible amount of energy at a given temperature.

Grey body : Grey body is defined as a body whose absorptivity is constant at all wavelengths of radiation, at a given temperature.

Emissivity : The energy emitted by actual body to the energy emitted by black body is known as emissivity.

Heat exchangers : Heat exchangers are the devices used for transferring heat from one fluid (hot gas or stream) to another fluid (liquid) through a metal wall.

Heat interchangers : Heat interchangers are the devices used for transferring heat from one liquid to another or from one gas to another gas through a metal wall.

Evaporation

Evaporation : Evaporation is a process of vaporizing large quantities of volatile liquid to get a concentrated product.

Or

The process of removal of solvent from the solution by boiling the liquid in a suitable vessel, and withdrawing the vapour, leaving a concentrated liquid residue in the vessel, is called evaporation.

Entrainment : The phenomenon in which finely divided liquid droplets are carried along with the stream of vapour is called entrainment.

Foam : Foam is the formation of a stable blanket of bubbles that lies on the surface of boiling milk (boiling liquids).

Calandria : Calandria is a steam compartment, which consists of a number of tubes fitted in a vessel and is included in the evaporator.

Size Reduction & Size Separation

Size reduction : Size reduction is a process of reducing large solid unit masses into small unit masses, coarse particles or fine particles. It is also termed as comminution or diminution or pulverization.

Cutting : Cutting is a means of tearing the material by a sharp blade.

Compression : Compression is a means by which the material is crushed between rollers by the application of pressure.

Attrition : Attrition involves breaking down of the material by rubbing action between two surfaces.

Impact : It involves the operation of splitting the material apart, when a lump of material strikes against the rotating hammers.

Size separation : Size separation is a unit operation that involves the separation of a mixture of various sizes of particles into two or more portions by means of screening surfaces. It also known as sieving / sifting / classifying / screening.

Coarse powder : A powder, all the particles of which pass through a sieve with nominal mesh aperture of 1.7 mm (No. 10 sieve) and not more than 40% through a sieve with nominal mesh aperture of 355 µm. (No. 44 sieve) is called coarse powder.

Sieve number : Sieve number indicates the number of meshes per linear length of 0.0254 m (one inch).

Blending : It means to mix smoothly and inseparably together.

Cetrifugation : It is a unit operation employed for separating the constituents present in dispersion with the aid of centrifugal force.

Actual Screen : It is the one that does not give perfect separation about the cut diameter.

Mixing

Mixing : Mixing is defined as a process that tends to result in a randomization of dissimilar particles with in a system

Connective mixing : It is achieved by the inversion of the powder bed using blades or paddles or screw element.

Shear mixing : In this type, the forces of attraction are broken down so that each particle moves on its own between regions of different composition and parallel to their surfaces.

Diffusive mixing : It involves the random motion of particles within the powder bed there by particles change their positions relative to one another.

Agitation : Agitation refers to the induced motion of a material in a specified way, usually in a circulatory pattern inside a container.

Bulk Transport : Bulk transport is defined as the movement of a large portion of a material from one location to another location in a given system.

Turbulent mixing : Turbulent mixing is defined as mixing due to turbulent flow, which results in random fluctuation of the fluid velocity at any given point within the system.

Laminar mixing : Laminar mixing is the mixing of two dissimilar liquids through laminar flow, i.e. the applied shear stretches the interface between them.

Molecular Diffusion : Molecular diffusion is the mixing at molecular level in which molecules diffuse due to thermal motion.

Pitch : Pitch is defined as the distance the impeller would move through the fluid per revolution, if slippage does not occur.

Filtration

Filtration : Filtration may be defined as a process of separation of solids from a fluid by passing the same through a porous medium that retains the solids, but allows the fluid to pass though.

Slurry : The suspension to be filtered is known as slurry.

Filter medium : The porous medium used to retain the solids is known as filter medium.

Filter cake : The accumulated solids on the filter are referred to as filter cake.

Filtrate : The clear liquid passing through the filter is filtrate.

Surface Filtration : It is a screening action by which pores or holes of the medium prevent the passage of solids. It is also known as screen filtration.

Depth Filtration : In this process, slurry penetrates to a point where the diameter of solid particles is greater than that of the void or channel.

Cake Filtration : A filter consists of a coarse woven cloth through which a concentrated suspension of rigid particles is passed, so that they bridge the holes and form a bed.

Filter Aid : It forms a surface deposit, which screens out the solids and also prevents plugging of the supporting filter medium.

Clarification : It is the process of separation of low concentration of solids (<1% w/v) from liquid.

Viva Questions

I Radiation Constant of Iron Cylinder

II Radiation Constant of Brass Cylinder

III Radiation Constant of Copper Cylinder

IV Painted Glass

V Unpainted Glass

1. What do you mean by radiation constant?
2. What do you mean by Conduction, Convection & Radiation?
3. What is Stephan Boltzman constant?
4. What is black body?
5. Why do you prefer Iron, Brass & Copper cylinders for determining Radiation Constant?
6. What is your intention to do this experiment?
7. How the temperature will influence on Radiation Constant?
8. What are incandescent solids?
9. In how many ways heat can be transferred to the metal walls?
10. What are the Pharmaceutical applications of Radiation Constant?

VI Size Reduction & Seperation

1. What are the synonyms for size reduction?
2. Why size reduction is extensively used in Pharmaceutical operations?
3. What are the mechanisms involved in size reduction?
4. What are the applications of size reduction?
5. What are the objectives of size reduction?
6. What is the principle involved in ball mill?
7. Why do you get different sizes of particles?
8. Why do you prefer same size of Balls?

VII Flow of Fluids

1. What is streamline flow & turbulent flow?
2. What do you mean by Reynolds Number?
3. What is critical velocity?
4. What are the factors that affect the fluid flow?
5. What are the limitations of Reynolds Number?
6. What is the significance of Reynolds Number?
7. How does viscosity of liquid, diameter & length of the pipe affect the friction?
8. In which flow Eddies are formed?

VIII Drying

1. What is the mechanism involved in Drying operation?
2. What are the different stages of drying?
3. What is critical moisture content?
4. What do you mean by first falling period?
5. What are the factors that affect the rate of Drying?
6. What is the importance of Drying?
7. What are the Pharmaceutical applications of drying?
8. What is equilibrium moisture content?

IX Particle Size Analysis

1. What are the different types of powders?
2. What is the type of arrangement of sieves during sieve analysis?
3. What do mean by tailing?
4. What are fines & tails?
5. What are least & maximum size of particles Pharmaceutically?
6. Why different sizes of particles used in preparation of Tablets?
7. In which area of pharmaceutical operations sieves analysis is used?
8. What are the specifications of standard sieves?

X Filtration

1. What is filter aid & give some examples?
2. What do you mean by filtration?
3. What are the factors affecting the rate of filtration?
4. How does the particle size affect the rate of filtration?
5. What is the equation that governs the effect of particle size on rate of filtration?
6. In how many ways filter aids can be used?
7. What are the applications of filtration in Industrial Pharmacy?
8. What are the different types of Filtration mechanisms?

XI Humidity

1. What is Humidity?
2. What is Dry bulb & Wet bulb temperatures?
3. What is adiabatic temperature?
4. What are adiabatic cooling lines?
5. What relative Humidity?
6. What do you mean by Dew point?
7. What is the use of Humidity chart?
8. What are the different methods to find Humidity?

XII Crystallization

1. What are the operations involved in crystallization?
2. What are the different ways to achieve super saturation?
3. What is salting out method?
4. What do you mean by adiabatic evaporation?
5. How the nuclei will form?
6. What do you mean the term Crystal growth?
7. What are the factors that affect the crystallization?
8. What is the effect of pH on crystallization?
9. What is solubility?
10. What effect of temperature on solubility?

XIII Distillation

1. What are the applications of distillation in Pharmacy?
2. What do you mean by vapour pressure?
3. What do you mean by boiling point?
4. What is relative volatility?
5. Classify the types of distillations?
6. What is steam distillation?
7. What do you mean Azeotrope (Entrainers)?
8. What are the applications of Azeotropic distillation?
9. How can prepare the aromatic water?
10. What are different types of azeotropic mixtures & give examples?
11. What is use of boiling point diagram?
12. What are colligative properties?

XIV Stoke's Law

1. What is stoke's law?
2. What are the factors affecting the stoke's law
3. What are the limitations of stoke's law
4. Name some Pharmaceutical dosage forms where stoke's law is applicable?
5. What do you mean by Terminal settle velocity?
6. What are the applications of stoke's law?
7. What is viscosity?
8. What is the viscosity of glycerin?

XV Centrifugation

1. What is Centrifugation?
2. What do you mean by Centrifugal force?
3. What do you mean by driving force?
4. What are the different types of Centrifuges?
5. What are the factors that affect the Centrifuges?
6. What are suspending agents & give examples?

XVI Extraction

1. What is maceration?
2. What is percolation?
3. What do you mean by tinctures & galanicals?
4. Enumerate different methods of maceration?
5. Enumerate different methods of percolation?
6. What do you mean Menstruum?
7. What is marc?
8. What is digestion?
9. What is infusion?
10. What is decoction?
11. What is effect of temperature on extraction?